Literature-Based Activities for Integrating Mathematics with Other Content Areas, Grades 6–8

Robin A. Ward
RICE UNIVERSITY

PEARSON

Boston New York San Francisco
Mexico City Montreal Toronto London Madrid Munich Paris
Hong Kong Singapore Tokyo Cape Town Sydney

Series Editor: Kelly Villella Canton
Series Editorial Assistant: Christine Pratt Swayne
Marketing Manager: Danae April
Production Editor: Paula Carroll
Editorial Production Service: Kathy Smith
Composition Buyer: Linda Cox
Manufacturing Buyer: Linda Morris
Interior Design and Composition Services: Schneck-DePippo Graphics
Cover Administrator: Linda Knowles
Cover Designer: Elena Sidorova
Director of Professional Development: Alison Maloney

For related titles and support materials, visit our online catalog at www.pearsonhighered.com.

Between the time website information is gathered and then published, it is not unusual for some sites to have closed. Also, the transcription of URLs can result in typographical errors. The publisher would appreciate notification where these errors occur so that they may be corrected in subsequent editions.

Cataloging in Publication Data on file at the Library of Congress

ISBN–13: 978-0-205-52916-2
ISBN–10: 0-205-52916-X

Printed in the United States of America

10 9 8 7 6 5 4 3 2 1 11 10 09 08

Allyn & Bacon
is an imprint of

PEARSON

www.pearsonhighered.com

Contents

About the Author

After earning a Bachelor's degree in math/physics from Immaculata College and a Master's of Arts degree in mathematics from Villanova University, Robin worked as an aerospace engineer and systems programmer. Prompted by a growing interest in teaching mathematics she pursued a PhD in mathematics education from the University of Virginia. Since graduating with her doctorate, Robin has spent ten years working as a professor of mathematics education at California Polytechnic State University and the University of Arizona. A devoted mathematics teacher educator, she sought and received numerous grants from NASA, NSF, and the U.S. Department of Education, all geared toward improving K–12 teachers' mathematical and pedagogical knowledge. A two-time recipient of a Stanford-ASEE fellowship, she partnered with the NASA Dryden Flight Research Center to develop web-based materials for K–12 teachers that showcase the work of NASA scientists. In 2005, the University of Arizona's College of Education bestowed upon her the Outstanding Teaching Award. Robin has written numerous articles and has presented many professional development workshops and talks at national conferences on improving mathematics pedagogy and using children's literature to enhance the teaching and learning of K–8 mathematics.

Robin lives in Houston with her husband, Chris DelConte, and two young daughters, Sienna and Sophia Arabella. There she pursues her passion for improving K–12 teachers' mathematical and pedagogical knowledge as the Associate Director of Curriculum Integration at the Rice University School Mathematics Project. In addition, she gives professional development workshops for local teachers and integrated mathematics-art classes for Pre-Kindergarten–Grade 8 students.

Acknowledgments

I would like to thank the many middle school teachers whose ideas served as the basis for several of the integrated activities included in this book. I also thank my former students who allowed me to use the classroom as a laboratory for my ideas on how to integrate children's literature into the teaching and learning of mathematics. Their active participation in literature-based classroom activities and their candid responses shared in their reflections assisted me in formulating and fine-tuning many of the ideas presented in this book.

Also, I'd like to acknowledge and thank Dawn Corso, my dear friend and colleague, for her insight and spirited efforts during the initial design phases of this book.

I am also grateful for the thoughtful comments and feedback on this book series from the reviewers lined up by Allyn and Bacon: Tammy Brown, Denver Public Schools; Karen Caldwell, John Early Paideia Middle School; Linda Cole, Barrie School; Bethany Dannelly, Hampton Roads Academy; Amanda Guinn, Monroe County Community Schools-University School; Cheri Howard, Fairbanks North Star Borough School District; Tim Linnet, Maywood Middle School; Kris O'Clair, Denver Public Schools; Lorel Preston, Westminster College; E. Elaine Rafferty, Charleston County School District; Ellen Szecsy, Math Consultant; Alexandra Thompson, Devon Elementary School; Karen Ward, Centennial Middle School.

Thank you to my other friends and colleagues spread far and wide, who supported me in a variety of immeasurable ways during this book endeavor and who continue to inspire me, namely, Gina Bernitt, Michelle Bickman, the Chan family, Elaine Chin, Patricia Reedy-Deserio, Winnie Doyle, Shirley Fisher,

Gassia Gerges, Mary Beth Gilbert, Therese Grahn, my friends at JMH, Cynthia Johnson, Ingrid Johnson, the Kinerk family, Lori Levitt, the Livengood family, Michelle Lopez-Varma, Michelle Lozano, Sheryl Maxwell, Tom McDuffie, Nicole O'Fiesh, Sheryl O'Neill, Anne Papakonstantinou and the RUSMP staff, Diana Perdue, Susan Roper, Gazala Siddiqui, Kim Thoss, Sheila Tobias, and Fred and Nancy Utter.

Finally, a huge thank you to my husband, parents, family, and two daughters. Sienna and Sophia Arabella, Mommy is all done!

Literature-Based Activities for Integrating Mathematics with Other Content Areas, Grades 6–8

Introduction

Reading in the mathematics classrooms? Absolutely! However, integrating literature into the teaching and learning of mathematics is more than just reading a book to students. By exploring and reading works of fiction, nonfiction, and poetry, students can engage in worthwhile and stimulating mathematical activities that encourage them to communicate their ideas verbally or through drawing or writing. In short, mathematics can be viewed as "a vehicle for thinking, a medium for creating, and a language for communicating" (Kleiman, 1991, p. 48). Additionally, using literature requires students to listen and comprehend—two vital skills needed for academic success. Thus, the goal of integrating literature into the teaching and learning of mathematics is to improve the overall literacy of students.

Using Literature to Teach Mathematics

Integrating literature into the teaching and learning of K–8 mathematics is gaining momentum. In fact, a growing body of research and anecdotal evidence documents the potential and power of using literature in mathematics classrooms (Avery & Avery, 2001; Burns, 1995; Burns & Sheffield, 2004; Capraro & Capraro, 2006; Carr, Buchanan, Wentz, Weiss, & Brant, 2001; Draper, 2002; Hellwig, Monroe, & Jacobs, 2000; Hunsader, 2004; Johnson & Giorgis, 2001; Leitze, 1997; Leu, Castek, Henry, Coiro, & McMullan, 2004; MacGregor & Price, 1999; Monroe & Livingston, 2002; Moyer, 2000; Schiro, 1997; Ward,

2003, 2004a, 2004b, 2004c, 2005, 2006a, 2006b; Ward & Muller, 2006; Whitin & Whitin, 1996, 2004; Whitin & Wilde, 1992, 1995; Young, 2001). Because many mathematical ideas and concepts are abstract or symbolic in nature, literature offers teachers the opportunity to present and discuss these ideas and concepts within the context of a story, using illustrations, prose, and more informal, familiar language. This, in turn, can make the learning of mathematics less intimidating and more engaging, especially for students whose first language is not English. Further, using literature to teach mathematics provides students with additional opportunities, encouragement, and support for speaking, writing, reading, and listening in mathematics classes.

Johnson and Giorgis (2001) contend that interacting with literature "encourages delight, nurtures curiosity and wonderment, and invites readers to discover things never experienced before" (p. 204). Because literature can be used to initiate a discussion on a social issue, introduce a theme, trigger a round of creative writing, or strengthen an appreciation for poetry, even secondary teachers are discovering the therapeutic and instructional value of this genre (Avery & Avery, 2001). In fact, it has been noted that picture books used in secondary mathematics and science curricula that relate content to the real world are beneficial for students' understanding of specific concepts and may encourage them to seek a career in the sciences (Carr et al., 2001). Additionally, many pieces of literature can be appreciated at different grade levels and can provide strategic opportunities for students to engage in problem solving.

Many national educational organizations, such as the National Council of Teachers of Mathematics (NCTM), the National Council of Teachers of English (NCTE), and the International Reading Association (IRA) also advocate that PreK–8 mathematics teachers regularly integrate literature into their teaching because of its many benefits. For example, in its *Principles and Standards for School Mathematics* (NCTM, 2000), NCTM asserts that "students who have opportunities, encouragement, and support for speaking, writing, reading, and listening in mathematics classes reap dual benefits: they communicate to learn mathematics, and they learn to communicate mathematically" (p. 60). NCTE (NCTE & IRA, 1996), noted for its history of commitment to the teaching of literature in connection with the teaching of reading and English/language arts, advocates the extensive use of children's and adolescent literature throughout the reading and writing curriculum and as a valuable source in content-area studies. It further endorses preparing and certifying teachers with strong content and pedagogical knowledge of children's and/or adolescent literature. Finally, in its position statement, IRA (2006) recognizes excellent reading teachers as those who are familiar with children's literature and who include a wide variety of fiction and nonfiction genres (e.g., storybooks, novels, and biographies) in their teaching.

Teaching Using an Integrated Approach

Gaining equal momentum is the movement by PreK–8 teachers to integrate their teaching across content areas with the goal of building connections between and among the various subject matters taught in school. Organizations such as NCTM, the National Council for the Social Studies (NCSS), the National Research Council (NRC), and the Music Educators National Conference (MENC) support an integrated approach to teaching. For example, in its *Curriculum Standards for Social Studies,* NCSS (1994) states that "a social studies perspective is academically sound, multidisciplinary, and integrative" (pp. xvii–xviii). The *National Science Education Standards* (NRC, 1996) offer that "school science and mathematics programs should be coordinated so that students learn the necessary mathematical skills and concepts before and during their use in the science program" (p. 214). Similarly, MENC (1994) has identified as its goal for the *National Standards for Arts Education* to "help students make connections between concepts and across subjects," as the learning tasks defined by these *Standards* serve as "bridges among the arts disciplines, and finally as gateways from the arts to other areas of study" (p. 13). NCTM (2000) echoes these same sentiments and advocates that students' mathematical experiences at all levels include opportunities for connections to other subject areas and disciplines, especially because mathematics permeates music and art content, and is used in science, the social studies, and other fields of study. Finally, the National Middle School Association (NMSA, 1995) argues that the school curriculum needs to be integrative in order to help young adolescents make sense out of their life experiences and connect school experiences to their daily lives outside of the classroom.

The benefits of an integrated curriculum, which recognizes that the subjects within the curriculum are connected to each other and to the real world, have been noted by educational philosophers, curriculum theorists, and others (Beane, 1993, 1995, 1997; Bruner, 1977; Dewey, 1924, 1933; Drake & Burns, 2004; Gelineau, 2003; Howey, 1996; Jacobs, 1989; Kim, Andrews, & Carr, 2004; McDonald & Fisher, 2006; Schwartz & Pollishuke, 2005; Vars, 1997; Wortham, 1996). Beane (1997) notes that for many students, the separate subject approach offers a "disconnected and incoherent assortment of facts and skills. There is no unity, no real sense to it all" (p. 42). Gardner (1997) also recognizes that students need to explicitly see connections and nonconnections between domains of knowledge; otherwise, generalization and transfer will not occur. Beane (1993, 1995, 1997) as well as others (Jacobs, 1989; Vars, 1997) advocate for integrated curriculum models because they center on the problems and interests of young adolescent learners and serve as a valuable lens for understanding student thinking (Perkins, 1989). According to Wortham (1996), "An integrated curriculum crosses subject areas . . . the intent is to construct meaningful bridges to show connections in development and learning" (p. 330). Thus, curriculum integration is often achieved through the design of integrated thematic units or through the study of a topic.

A growing body of literature continues to document the effectiveness of integrated curriculum on student achievement (Arhar, 1997; Cornett, 2003; Davies, 1992; Drake, 1998; Drake & Burns, 2004; National Association for Core Curriculum, 2000; Vars, 1996, 1997; Vars & Beane, 2000; Watts, 2004) and its ability to increase student motivation, elicit higher order thinking, and build stronger interpersonal skills (Vars, 1997). Bailey (2000), Caskey (2001), and Caskey and Johnston (1996) argue that an integrated curriculum is warranted, as a truly integrated curriculum enables teachers and their students to make connections between real life and their classroom learning experiences. Further, an integrated curriculum engages teachers, stimulates students, and energizes classroom learning environments (Meinbach, Fredericks, & Rothlein, 2000).

Teaching Collaboratively

Although teaching middle school is often characterized as an isolated activity, a vibrant shift in teaching is taking place whereby teachers are participating in "communities of practice" (Lave & Wenger, 1991) during which they assess student understanding; plan, design, and implement new instructional ideas; and reflect upon their teaching (Franke & Kazemi, 2001; Kazemi & Franke, 2004; Lieberman & Grolnick, 1997; Louis & Marks, 1998; Palincsar, Magnusson, Marano, Ford, & Brown, 1998). Opportunities for teacher collaboration allow for teachers to articulate different pedagogical styles, express differing beliefs and perspectives, agree and disagree, and weigh alternatives (Hindin, Morocco, Mott, & Aguilar, 2007). Further, teachers who engage in communities of practice can improve their classroom practices and share their various fields of expertise, which, in turn, can stimulate and renew their intellectual growth, particularly in their content-area knowledge (Grossman, Wineburg, & Woolworth, 2001).

Many national organizations strongly advocate teacher collaboration in and across content areas, in particular, at the middle and high school levels. NCTM (2000) recommends that teachers collaborate with colleagues regularly to "observe, analyze, and discuss teaching and students' thinking" (p. 19) and that teachers turn to colleagues in order to "expand their views of mathematics, their resources for teaching, and their repertoire of teaching and learning skills" (NCTM, 1991, p. 169). NRC (1996) asserts that teachers "work together as colleagues within and across disciplines and grade levels" as "individual and collective planning is a cornerstone to science teaching" (p. 32). Similarly, in its *Curriculum Standards for the Social Studies,* NCSS (1994) recognizes the need for "collegial planning time" (p. 174), given the integrative nature of the social studies curriculum, as does MENC (1994), citing that "the intellectual methods of the arts are precisely those used to transform scientific discoveries and discoveries into everyday technology" (p. 132).

Using Literature Combined with an Integrated and Collaborative Approach to Teaching

Given that more and more teachers are turning to literature as a means to enliven and demystify the study of mathematics, and given that teachers are striving to build rich connections for students by using an integrated and collaborative approach to teaching, this book was written to address these merging efforts. This book series is divided into three volumes for the following grade bands: K–2, 3–5, and 6–8. Within each of these grade-specific volumes are three chapters of classroom-tested activities. The first chapter in each series articulates literature-based activities that integrate mathematics and science; Chapter 2 contains literature-based activities that integrate mathematics and the social studies; and Chapter 3 describes literature-based activities that integrate mathematics and the visual arts.

Each of the chapters opens with a brief overview, articulating the connection between mathematics and the specific content area (whether it be science, social studies, or the visual arts), followed by a list of the concepts and skills featured in the literature-based activities. A matrix follows, which lists each piece of literature used in that chapter's integrated activities and identifies other relevant cross-curricular concepts and skills. The collection of literature-based activities follows next.

Featured in many of the literature-based activities in this book are book pairs and, in some cases, book trios: that is, two or three pieces of literature that work powerfully together when teaching an integrated lesson. Also, several biographies are featured as well as works of nonfiction and poetry.

The format of each activity is as follows:

Book Title (along with the name(s) of the author(s), publisher, and ISBN number)

Overview of Book (brief overview of the story line of each book or poem featured in the activities)

Mathematical Concepts and Skills (list of the mathematical concepts and skills serving as the focus of the activities)

Content Area Concepts and Skills (list of the content-area concepts and skills serving as the focus of the activities)

Overview of Activities (brief description of the activities to follow)

National Mathematics Standards (mathematical expectations of students as defined by NCTM)

National Standards for the Content Area (content-area expectations of students as defined by NRC, NCSS, or MENC)

Materials (list of necessary materials)

Description of Activities (step-by-step procedure describing how to implement the literature-based activities)

Assessment (considerations and questions to ask when assessing student understanding of the concepts and skills presented in activities; see the appendix for more assessment ideas and rubrics)

Activity Extensions (description of potential follow-up activities)

Cross-Curricular Connections (brief description of related follow-up activities for other content areas)

Related Literature (list of other pieces of literature that feature the same or similar concepts and skills presented in the activities)

Related Instructional Resources (list of resources useful to the teacher)

Related Websites (list of websites pertinent to the concepts and skills presented in the activities)

Worksheets/handouts (if applicable)

Located at the end of the book is an extensive bibliography of over 200 pieces of literature cited throughout this book as well as over 150 citations of instructional resources that will support the teacher. Additionally, an appendix includes several assessment tools and rubrics useful in evaluating students' performance, skills, and abilities as they engage in the literature-based activities.

Finally, prior to implementing the integrated, literature-based explorations articulated in this book, read through the entire activity *prior* to class in order to assess whether you need to modify or delete any steps in the activities due to student accommodations, lack or substitution of materials, limited resources or space, prior preparation of materials, and so on. Do not feel as though these activities are scripted! Use your judgment and pedagogical ingenuity to take tangents as deemed appropriate and to capitalize on those teachable moments, as the ultimate goal is to maximize student understanding and to make the teaching and learning of mathematics and the other content areas come to life. Also, consider maintaining each piece of literature that you use in the classroom library, allowing students access to the books should they want to view them later.

The author hopes that the interdisciplinary, literature-based activities presented in this book will provide readers with the information, resources, and confidence to make the teaching and exploration of mathematics, science, social studies, and the arts more meaningful, cohesive, interactive, and exciting to learners.

Regarding Assessment

Toward the end of each activity in this book, several questions are included under the heading, Assessment that a teacher might consider posing as a means to assess students' understanding of the concepts and mastery of the skills presented in each of the literature-based activities. NCTM (1989, 2000) advocates that student assessment be integral to instruction, teachers use multiple means of assessment, and that all aspects of mathematical knowledge (confidence, attitude, problem-solving abilities, etc.) be assessed. Examples of multiple means of assessment include observations, interviews and questions, student journal entries, portfolios, student self-evaluations, and peer evaluations.

Included in the appendix to this book are several assessment tools and rubrics a teacher might employ as a means to better assess students as they engage in the literature-based activities. Other helpful assessment resources are included at the end of this book in the Assessment Resources References section.

1

Literature-Based Mathematics and Science Activities

The Mathematics-Science Connection

Science encompasses many domains including life science, physical science, and earth and space science. It is the study and exploration of our world, our universe, our environment, and other phenomena. In studying science, students should not memorize facts but, instead, be encouraged to think, observe, hypothesize, reason, communicate, and problem solve, activities the NCTM (1989, 2000) and the NRC (1996) strongly advocate. Recognizing the interplay between mathematics and science, the German mathematician Carl Gauss once stated that "Mathematics is the Queen of the Sciences." Given the strong interconnectedness between mathematics and science, a mounting movement continues to support the integration of mathematics and

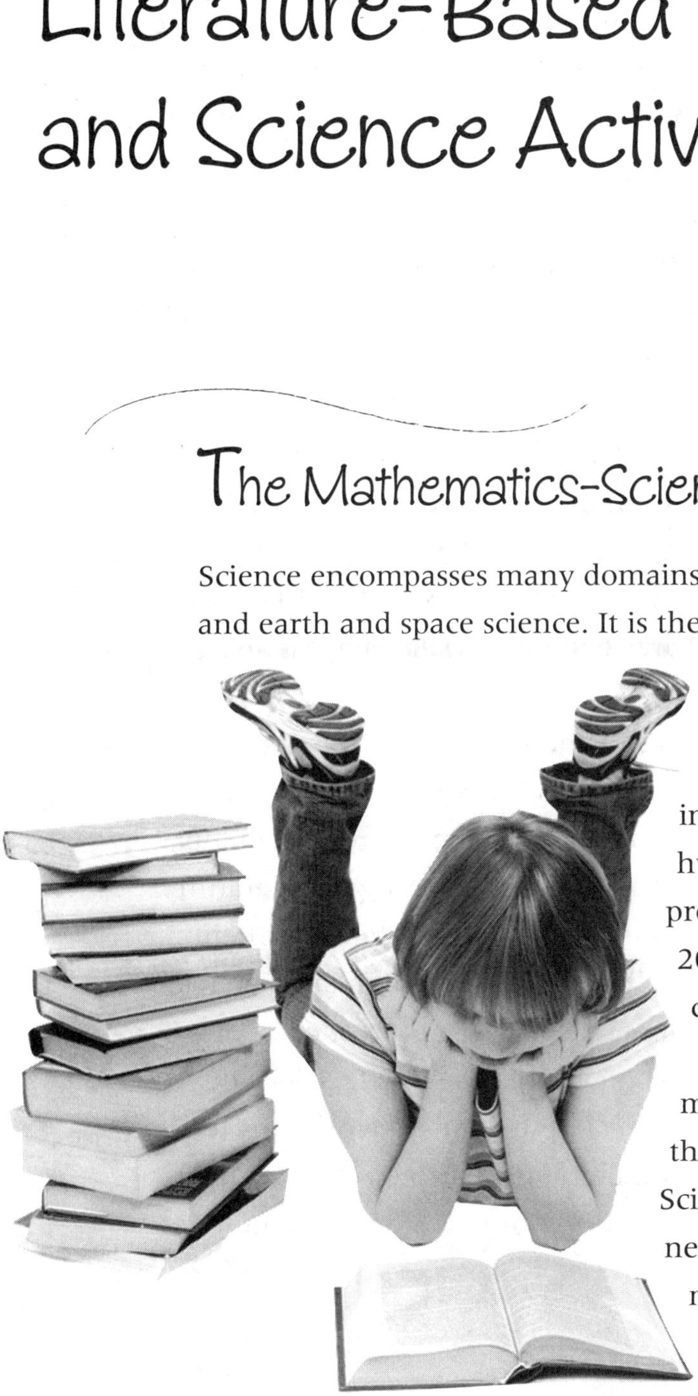

science in the classroom curriculum (Basista & Mathews, 2002; Cobb, 2000; Johnson & Giorgis, 2001, Kaser, 2001; Moyer, 2000; Putnam & Borko, 2000; Roth & McGinn, 1998). One of the best ways for students to increase their knowledge and understanding of their world in order to make sense out of nature's sometimes complicated phenomena and today's advanced technological society is to integrate literature into the study of science and mathematics.

This chapter articulates a variety of literature-based activities that integrate concepts and skills used and learned in the study of mathematics with those in science. While engaged in these activities, students will discover and gain practice with such mathematics concepts and skills as fractions, decimals, estimation, ratio, proportional reasoning (Number and Operations Standard); functions, linear equations, slope, y-intercept, algebra, solving for one unknown, exponential decay, exponential functions (Algebra Standard); circles, two- and three-dimensional shapes, lines, angles (Geometry Standard); size, scale, metric and customary units, diameter, area and circumference of a circle, angle measurement, period, frequency (Measurement Standard); and probability, randomness, data collection and interpretation, box and whiskers plot, average, mean, median, mode, Law of Large Numbers (Data Analysis and Probability Standard).

"Children's literature can help young learners make sense out of nature and today's advanced technological society."

Science concepts and skills featured in this chapter include investigations, formulating questions, gathering evidence, scientific inquiry (Science as Inquiry, Content Standard A); properties and changes of properties in matter, transfer of energy, motions and forces, measuring and recording temperature, Fahrenheit, Celsius, degrees, thermometer, scale (Physical Science, Content Standard B); structure and function in living systems, regulation and behavior of organisms, organisms' internal environment, survival and extinction of organisms, food web, diversity and adaptation of organisms (Life Science, Content Standard C); size and scale of planets in the solar system, characteristics of planets, orbit, axis, rotation, movement of the earth on its axis and about the sun, eclipses (Earth and Space Science, Content Standard D); understandings about science and technology (Science and Technology, Content Standard E); science and technology in society (Science in Personal and Social Perspectives, Content Standard F); and science as a human endeavor, history of science, inventions and inventors (History and Nature of Science, Content Standard G).

The integrated literature-based activities also provide students with many opportunities to predict, estimate, problem-solve, and reason (Problem Solving and Reasoning and Proof Standards) as well as communicate and use various representations to organize, record, model, and interpret mathematical ideas (Communication and Representations Standards). Further, students will discover and explore real-life applications of mathematics and science and careers in mathematics and science (Connections Standard).

Remember to check the appendix for ideas for assessment tools and rubrics.

Matrix of Mathematics and Science Activities

BOOK TITLE	MATHEMATICAL CONCEPTS AND SKILLS	SCIENCE CONCEPTS AND SKILLS	SOCIAL STUDIES CONCEPTS AND SKILLS	VISUAL ARTS CONCEPTS AND SKILLS
An Extraterrestrial Alien (a poem in *A Pizza the Size of the Sun*); *Our Earth and the Solar System*	fractions, decimals, scientific notation, measurement, ratio, size and scale, estimation	planets, size and scale of planets in the solar system, planets' composition, geology, atmosphere, orbit, and habitability	explore and discuss biographies of astronauts, scientists, or astronomers	explore and imitate artists' renditions of the night sky, space, or the solar system
Biggest, Strongest, Fastest	estimation, comparison of numerical quantities, measurement	animal characteristics and behavior, diversity and adaptation of organisms	research and discuss the implications of fast-growing populations on animal habitats and ecosystems	explore the works of wildlife artists
A Game of Functions; Temperature; How Little and How Much: A Book about Scales	functions, dependent and independent variables, linear equations, slope, y-intercept, algebra, solving for one unknown, data collection, graphing, measurement, using metric and customary units	measuring and recording temperature, Fahrenheit, Celsius, degrees, thermometer, scale	research and discuss demographics and geographical facts about very hot and very cold places	explore artwork depicting weather and the seasons
Starry Messenger; Galileo Galilei (Scientists Who Made History Series)	functions, dependence, period, frequency, measurement, graphing, prediction	scientific inquiry, motions and forces, history of science	discuss the impact of Galileo's work on the fields of science, astronomy, and philosophy	explore surrealism by viewing Salvador Dali's "melting" clocks
Great Buildings (Discoveries Series); *Building Big*	drawing and building two- and three-dimensional objects, measurement, ratio, proportion, scale	science and technology, contributions of technology	explore and discuss the need for laws and policies to govern scientific and technological applications	explore the use of the golden mean in architecture

BOOK TITLE	MATHEMATICAL CONCEPTS AND SKILLS	SCIENCE CONCEPTS AND SKILLS	SOCIAL STUDIES CONCEPTS AND SKILLS	VISUAL ARTS CONCEPTS AND SKILLS
A Pizza the Size of the Sun (a poem in *A Pizza the Size of the Sun*); *Eclipse: Darkness in Daytime*	circumference, diameter, radius, area of a circle, ratio and proportion, fractions, decimals, percents, angle measurement	solar eclipse; movement of and relationship between and among the earth, moon, and sun; corona; umbra; penumbra; Baily's Beads	research and discuss how ancient people interpreted and responded to eclipses	explore and imitate Van Gogh's *A Starry Night*, research and discuss the sun's role in architecture
Human Body (DK Guide Series)	measurement, ratio, proportional reasoning, estimation, data collection	structures and functions in living systems	explore and discuss a biography of a scientist	explore artists' use of the golden mean in architecture
Almost Gone: The World's Rarest Animals	estimation, data creation and interpretation, graphing, box and whiskers plot, average, mean, median, mode, range, extremes, lower quartile, upper quartile, outlier	animal habitats, regulation and behavior of organisms, organisms' internal environment, survival and extinction of organisms	locate animal habitats on a world map, research and discuss the implications of fast-growing populations on natural resources and food and water supplies	explore the works of wildlife artists
Marie Curie	prediction, probability, randomness, Law of Large Numbers, data collection, graphing, data interpretation and analysis, halving, fractions, exponential decay, exponential functions	properties and changes of properties in matter, transfer of energy, science and technology, science as a human endeavor, history of science	explore and discuss the impact and repercussions of dropping the atomic bombs, read and discuss a biography of Wilson "Snowflake" Bentley	explore the works of artists who experimented with light and color
I Made Something Strange with My Chemistry Set (a poem in *A Pizza the Size of the Sun*); *So You Want to Be an Inventor?*; *Perfect Inventions = Perfect World*	real-life applications of mathematics, careers in mathematics	science and technology, inventions and inventors, science as inquiry	explore and discuss the inventions of and discoveries made by Archimedes	research and discuss the sketches of Leonardo da Vinci's inventions, create a collage of inventors

Activities Featuring Number and Operations

"An Extraterrestrial Alien" (a poem in A Pizza the Size of the Sun) (1996)

by Jack Prelutsky

Greenwillow Books, ISBN #0590374699

Our Earth and the Solar System (2001)

by Ken Graun

Ken Press, ISBN #1928771025

Overview of Poem and Book: Find out what happens when an extraterrestrial alien appears in a child's room in the poem, "An Extraterrestrial Alien." Explore facts about the earth, sun, moon, planets, comets, asteroids, galaxies, and more in *Our Earth and the Solar System,* a fact-filled book about our solar system.

Mathematical Concepts and Skills: fractions, decimals, scientific notation, measurement, ratio, size and scale, estimation

Science Concepts and Skills: planets, size and scale of planets in the solar system, planets' composition, geology, atmosphere, orbit, and habitability

Overview of Activities: Students explore the size of planets in our solar system using Playdoh as a concrete model. Students also estimate, predict, and compute relative planet sizes and use scientific notation to express their sizes and distances from the sun. Students research and present facts about the planets (e.g., planets' composition, geology, atmosphere, orbit, and habitability) and offer conjectures about how gravity is the force that holds planets and planets' moons in their orbits.

National Mathematics Standards (2000): Students in grades 6–8 should "work flexibly with fractions, decimals, and percents to solve problems" and "develop an understanding of large numbers and recognize and appropriately use exponential, scientific, and calculator notation" (p. 393). Students should "understand and use ratios and proportions to represent quantitative relationships" (p. 393) (Number and Operations Standard).

National Science Standards (1996): Students in grades 5–8 should "develop an understanding of earth and the solar system as a set of closely coupled systems" (p. 158). Students should understand that "the earth is the third planet from the sun in a system that includes the moon, the sun, eight other planets and their moons, and smaller objects, such as asteroids and comets" (p. 160). Students should also understand that "most objects in the solar system are in regular and predictable motion" and that these motions "explain such phenomena as the day, year, phases of the moon, and eclipses" (p. 160). Students should understand that "gravity is the force that keeps planets in orbit around the sun" (p. 161) (Earth and Space Science, Content Standard D).

Materials: 5 pounds of Playdoh; plastic knife; nine pieces of paper, each one labeled with the name of a planet

Description of Activities:

1. Read the poem, "An Extraterrestrial Alien," by Jack Prelutsky to set the stage for the upcoming activities.
2. Students work in pairs and develop a list of all of the planets in our solar system in order from biggest to smallest. (This activity includes Pluto as a planet, even though it is now classified as a dwarf planet.) As a whole class, discuss, agree upon, and record on the board the ordering of the planets from biggest to smallest.
3. Divide the class into two groups. Students develop a scaled model of the planets in our solar system by continually dividing a large mound of Playdoh into fractional pieces and then shaping it into balls, forming planets. (This activity is a modified version of the Worlds in Comparison activity created by Project Astro.)
    a. Prior to the activity, print the name of each planet on a piece of paper. Place all nine pieces of paper on a long table.
    b. Roll 5 pounds of Playdoh into a log shape. (It is acceptable if the Playdoh is different colors since planets are multicolored.)
    c. Cut the Playdoh into ten equal pieces. Take six pieces and place them on the sheet of paper labeled *Jupiter*. Place three pieces on the paper labeled *Saturn*.
    d. Take the remaining one piece and roll it into a log and cut it again into ten equal pieces. Place five pieces on the paper labeled *Saturn*. Place two pieces on the paper labeled *Neptune*. Place two pieces on the paper labeled *Uranus*.
    e. Take the remaining one piece and cut it into four equal pieces. Take three pieces and place them on the paper labeled *Saturn*.

f. Cut the remaining one piece into ten equal pieces. Place two on the paper labeled *Earth*. Place two pieces on the paper labeled *Venus*. Place four pieces on the paper labeled *Uranus*.

g. Roll the remaining two pieces into a log and cut into ten equal pieces. Place one piece on the paper labeled *Mars*. Place four pieces on the paper labeled *Neptune*. Place four pieces on the paper labeled *Uranus*.

h. Cut the remaining one piece into ten equal pieces. Place seven on the paper labeled *Mercury*. Place two pieces on the paper labeled *Uranus*.

i. Cut the remaining one piece into ten equal pieces. Place nine on the paper labeled *Uranus*. Place one on the paper labeled *Pluto*.

j. For each planet, roll all of the pieces of Playdoh into a ball, forming nine individual planets.

4. With the help of the group, one student from each group slides the labeled pieces of paper around on the table and orders the planets from biggest to smallest in terms of the sizes of the balls of Playdoh. Compare this ordering to students' original predictions. How accurate were their predictions? Which planet is biggest? Smallest? About equal size? Are students surprised at how many times larger Jupiter is, for example, compared to Earth? To Mercury? To other planets? Compare and check for accuracy the students' size ordering of the planets by viewing the Order of Planets from Biggest to Smallest website, or view all the planets at once and see their relative sizes on the Relative Sizes of the Planets in the Solar System website.

5. Ask students to look at their Playdoh planets and to visually estimate and record as a ratio how many times larger one planet is relative to another (e.g., Earth looks approximately three times larger than Mercury, thus the ratio of Earth to Mercury is 3 to 1 or 3:1). Students compare their ratios to the actual sizes of planets using the information appearing in the orange-colored text boxes on each planet's page in *Our Earth and the Solar System*. Students might also refer to the Planet Size Comparison website listed below.

6. Read select passages from *Our Earth and the Solar System,* which provide additional information about the planets in our solar system. Consider reading the opening paragraph that briefly introduces each planet.

7. Students work in small groups and select (or are assigned) a planet. (Students might also explore the sun, the moon, asteroids, or comets.) Students research facts about their planet including its composition, geology, atmosphere, location in the solar system, habitability, and so on. Students record their planet's diameter and distance to the sun using scientific notation. Using the table appearing on the last page of *Our Earth and the Solar System,* students use ratios to compute (and record using scientific notation) the amount of time it would take to travel from the sun to their planet. This will give students an understanding of the scale of our solar system.
8. Students create a series of posters, booklet, slideshow presentation, or diorama displaying their findings. Students share their creations and facts with the class.
9. After students share their findings, students make conjectures about what is the force that holds planets in their orbits around the sun (gravity). Challenge students to articulate how a planet's gravitation field dictates the number of moons orbiting a planet.

Assessment:

- Did students make accurate predictions about the relative sizes of planets?
- Did students make good visual estimations of the comparative sizes of planets?
- Did students accurately compare the size of planets using ratios?
- Did students collect and present accurate findings about their planet?
- Did students correctly express their findings using scientific notation?
- Did students make reasonable conjectures about gravity, how it holds planets in their orbits, and how gravity explains why some planets have a greater number of moons as compared to others?

Activity Extensions:

- Challenge students to estimate how many times bigger the sun is compared to Earth and how many times smaller the moon is as compared to the earth. Students then view the Planet Size Comparison website to check their predictions. Are students surprised to learn how massive the sun is as compared to the earth as well as other planets? Were their predictions accurate regarding how many times larger the earth is as compared to the moon?
- Visit the Toilet Paper—Planets' Distances from the Sun website or the School Yard Solar System website for ideas on how students can create a scaled model of the universe, representing the planets' distances from the sun.

- Students use the Your Age on Other Worlds website or the Your Weight on Other Planets website listed below and discover how old they would be and how much they would weigh if they lived on another planet in our solar system. Students then compute the scale factor used for each planet to convert ages and weights.

Cross-Curricular Connections:

Social Studies

- Explore and discuss biographies of astronauts (Buzz Aldrin, Neil Armstrong, Sally Ride, etc.), astronomers, or scientists who studied bodies in our solar system (Copernicus, Galileo, Johannes Kepler, Edwin Hubble, Thomas Bopp, Carl Sagan, etc.).

Visual Arts

- Explore and imitate artists' renditions of the night sky, space, or the solar system (e.g., Van Gogh's *The Starry Night*).

Related Literature

Branley, F. (1998). *The planets in our solar system*. New York: HarperCollins.

Campbell, J., & Collison, C. (2005). *G is for galaxy: An out of this world alphabet*. Chelsea, MI: Sleeping Bear Press.

Gibbons, G. (2005). *Planets*. New York: Holiday House.

Goldsmith, M. (2001). *Galileo Galilei* (Scientists who made history series). New York: Raintree Steck-Vaughn.

Graun, K. (2001). *Our earth and the solar system*. Tucson: Ken Press.

Kerrod, R. (2003). *Universe* (Eyewitness books series). New York: Dorling Kindersley.

Levy, D. (2003). *Stars and planets* (Discoveries series). New York: Barnes & Noble Books.

Prelutsky, J. (1996). *A pizza the size of the sun*. New York: Greenwillow Books.

Scieszka, J., & Smith, L. (1995). *Math curse*. New York: Penguin Books.

Scieszka, J., & Smith, L. (2004). *Science verse*. New York: Penguin Books.

Sis, P. (1996). *Starry messenger*. New York: Farrar Straus Giroux.

Related Instructional Resources

Becklake, S. (1993). *The visual dictionary of the universe* (Eyewitness visual dictionaries series). New York: Dorling Kindersley.

Dickinson, T. (1995). *Other worlds: A beginner's guide to planets and moons*. Tonawanda, NY: Firefly.

Furniss, T. (1999). *Atlas of space exploration*. Milwaukee: Gareth Stevens.

Holland, S. (2001). *Space* (Eye wonder series). New York: Dorling Kindersley.

Schwartz, D. (2001). *Q is for quark: A science alphabet*. Berkeley, CA: Ten Speed Press.

Simon, S. (1992). *Our solar system, vol. 1*. New York: HarperCollins.

Simon, S. (1995). *The solar system: Facts and exploration*. New York: Holt.

Staff of Barron's. (2004). *Our solar system: A first introduction to space and the planets*. Hauppauge, NY: Barron's Educational Series.

Stott, C., & Twist, C. (2002). *1,001 facts about space*. New York: Dorling Kindersley.

Van Rose, S. (2005). *Earth* (Eyewitness books series). New York: Dorling Kindersley.

Related Websites

Family Astro
http://www.astrosociety.org/education/family.html

Hands-On Astronomy Activities
http://www.astrosociety.org/education/activities/handson.html
http://www.astrosociety.org/education/activities/astroacts.html

Order of Planets from Biggest to Smallest
http://coolcosmos.ipac.caltech.edu/cosmic_kids/AskKids/planet_sizes.shtml

The Planets
http://www.nineplanets.org/
http://solarsystem.jpl.nasa.gov/planets/profile.cfm?Object=SolarSys&Display=Overview

Planet Size Comparison
http://www.sciencenetlinks.com/interactives/messenger/psc/PlanetSize.html

Playdoh Recipe
http://www.kinderplanet.com/playdo.htm

Project Astro
http://www.astrosociety.org/education/astro/project_astro.html

Relative Sizes of the Planets in the Solar System
http://www.solarviews.com/raw/misc/ss.gif

School Yard Solar System
http://nssdc.gsfc.nasa.gov/planetary/education/schoolyard_ss/

Surfing the Solar System: A Treasure Hunt Game and Puzzle
http://www.astrosociety.org/education/surf.html

Toilet Paper—Planets' Distances from the Sun
http://www.astrosociety.org/education/activities/handson.html

Views of the Solar System
http://www.solarviews.com/eng/homepage.htm

Women in Astronomy
http://www.astrosociety.org/education/resources/womenast_bib.html

Your Age on Other Worlds
http://www.nasa.gov/audience/forkids/games/CS_Age_on_Saturn.html http://www.exploratorium.edu/ronh/age/

Your Weight on Other Planets
http://www1.nasa.gov/audience/forkids/games/A_Weight_on_Mars.html
http://www.exploratorium.edu/ronh/weight/

Biggest, Strongest, Fastest (1995)

by Steve Jenkins

Houghton Mifflin, ISBN #0395861365

Overview of Book: Learn the names of and fascinating facts about animals that hold the animal world record for being the biggest, smallest, fastest, slowest, strongest, and more.

Mathematical Concepts and Skills: estimation, comparison of numerical quantities, measurement

Science Concepts and Skills: animal characteristics and behavior, diversity and adaptation of organisms

Overview of Activities: Students learn facts about which animals are the biggest, smallest, fastest, slowest, strongest, and longest animals. Knowing the cheetah is the fastest species in the animal kingdom, students estimate the speeds of various animals, vehicles, people, and things. Students then work in pairs to research and present information about the various animals featured in *Biggest, Strongest, Fastest,* including a description of the animal's habitat, characteristics of the species, unique characteristics resulting from adaptation and evolution, food web, possibility of extinction, and so on.

National Mathematics Standards (2000): Students in grades 6–8 should "compute fluently and make reasonable estimates" and "develop meaning for integers and represent and compare quantities with them" (p. 393) (Number and Operations Standard). Students in grades 6–8 should "analyze and evaluate the mathematical thinking and strategies of others" (p. 402) (Communication Standard).

National Science Standards (1996): Students in grades 5–8 should develop an understanding that "food webs identify the relationships among producers, consumers, and decomposers in an ecosystem" and that "species acquire many of their unique characteristics through biological adaptation," which include "changes in structures, behaviors, or physiology that enhance survival and reproductive success in a particular environment" (p. 158). Students in grades 5–8 should develop an understanding that "an organism's behavior evolves through adaptation to its environment" and that "how a species moves, obtains food, reproduces, and responds to danger are based in the species' evolutionary history" (p. 157) (Life Science, Content Standard C).

Materials: *Biggest, Strongest, Fastest* worksheet, *Biggest, Strongest, Fastest* answer key

Description of Activities:

1. Challenge students to name which animal they believe to be the biggest, strongest, and fastest in the animal kingdom and to defend their reasoning. Begin a brief discussion about how species have evolved such that they have become the biggest, strongest, and fastest.
2. Read *Biggest, Strongest, Fastest.* After reading the text that states which animal is the biggest (or fastest, or longest, etc.) let students guess just how big (or how fast, or how long, etc.) *before* reading the smaller accompanying text that specifies the statistic that earns the animal its title.
3. Provide students with a copy of the *Biggest, Strongest, Fastest* worksheet. Students work in pairs to estimate (and then record in the first column) the speed of each of the items listed. Students then rank (and record in the second column) the items in order from slowest to fastest.
4. Prior to sharing the solutions, students as a class discuss their estimations, rankings, and defend their reasoning.
5. Provide students with the answers using the *Biggest, Strongest, Fastest* answer key. Students should draw comparisons between and among the speeds of various items on the list and express them using ratios, giving their speeds more meaning (e.g., The ratio of the speed of a cheetah at top speed to a car traveling on a freeway is approximately 1 to 1 or 1:1. This means that both objects travel at the same rate. The ratio of a commercial airliner at cruising altitude to the fastest official lap speed at Indy 500 is approximately 2 to 1 or 2:1. This means that the airliner travels twice as fast as the fastest Indy car.).
6. Place students in pairs and assign each student an animal from Jenkins's book. Students research facts about each animal, including a description of the animal's habitat, characteristics of the species, unique characteristics resulting from adaptation and evolution, food web, possibility of extinction, and so on, and present their findings to the class. Students present their findings as a poster, diorama, or a slideshow presentation.

Assessment:

- Did students make reasonable estimations for the speeds of the various items?
- Did students clearly justify their thinking in support of their estimations?

- Did students correctly express the speed ratios of various items on the list?
- Did students present accurate findings regarding their animal?

Activity Extensions:

- Read *Almost Gone* (Jenkins, 2006) and learn about which animals are extinct or endangered, and why.
- Read and enjoy poems about size and scale including "Whopper" (Prelutsky, 1996) and "It's Hard to be an Elephant" (Prelutsky, 1996).

Cross-Curricular Connections:

Social Studies

- Research and discuss the implications of fast-growing populations on animal habitats and ecosystems.

Visual Arts

- Explore the works of wildlife artists, such as John James Audubon.

Related Literature

Freedman, R. (1981). *Animal superstars: Biggest, strongest, fastest, smartest.* Englewood Cliffs, NJ: Prentice-Hall.

Jenkins, S. (1995). *Biggest, strongest, fastest.* Boston: Houghton Mifflin.

Jenkins, S. (2004). *Actual size.* Boston: Houghton Mifflin.

Jenkins, S. (2005). *Prehistoric actual size.* Boston: Houghton Mifflin.

Jenkins, S. (2006). *Almost gone: The world's rarest animals.* New York: Scholastic.

Jenkins, S., & Page, R. (2003). *What do you do with a tail like this?* Boston: Houghton Mifflin.

Prelutsky, J. (1996). *A pizza the size of the sun.* New York: Greenwillow Books.

Scieszka, J., & Smith, L. (1995). *Math curse.* New York: Penguin Books.

Scieszka, J., & Smith, L. (2004). *Science verse.* New York: Penguin Books.

Related Instructional Resources

Burnie, D. (2004). *Bird* (Eyewitness books series). New York: Dorling Kindersley.

Clark, B., & Buller, L. (2005). *Amphibian* (Eyewitness books series). New York: Dorling Kindersley.

Hare, T. (2005). *Animal fact file: Head-to-tail profiles of more than 90 mammals.* New York: Checkmark Books.

Mound, L. (2004). *Insects* (Eyewitness books series). New York: Dorling Kindersley.

National Geographic Society. (2000). *National Geographic animal encyclopedia.* Hanover, PA: National Geographic Children's Books.

Parker, S. (2004). *Mammal* (Eyewitness books series). New York: Dorling Kindersley.

Parker, S. (2005). *Fish* (Eyewitness books series). New York: Dorling Kindersley.

Schwartz, D. (2001). *Q is for quark: A science alphabet.* Berkeley, CA: Ten Speed Press.

Twist, C. (2005). *Reptiles and amphibians dictionary: An A to Z of cold-blooded creatures.* New York: Scholastic.

Related Websites

Animal Facts and Statistics

http://www.indianchild.com/amazing_facts.htm
http://lsb.syr.edu/projects/cyberzoo/a_list.html
http://www.factmonster.com/ipka/A0768508.html
http://sunsite.berkeley.edu/KidsClick!/midanim.html
http://www.kidsplanet.org/factsheets/map.html
http://www.sandiegozoo.org/animalbytes/index.html
http://www.worldalmanacforkids.com/explore/animals.html

Record Holders

http://www.guinnessworldrecords.com/
http://www.recordholders.org/en/links.html

Biggest, Strongest, Fastest Worksheet

Name(s): ____

Name of Object	My Estimation (mph)	Rank (Slowest to Fastest)	Actual Speed (mph)
Cheetah			60
Secretariat, world's fastest Triple Crown winner			
Commercial airline at cruising altitude			
Hurricane force wind			
Speeding bullet			
Carl Lewis, Olympic gold medalist in 100m and 200m track and field			
Fastest official lap speed at Indy 500			
Snail crawling			
F/A-16 fighter jet			
Titanic at impact with iceberg			
Speed of sound			
World's fastest roller coaster			
Car backing out of a driveway			
World's fastest elevator			
Maximum legal speed on a U.S. freeway (may vary from state to state)			

Biggest, Strongest, Fastest Answer Key

Name of Object	Rank (Slowest to Fastest)	Actual Speed (mph)
Snail crawling	1	0.03
Car backing out of a driveway	2	3
Carl Lewis, Olympic gold medalist in 100m and 200m track and field	3	22.5
Titanic at impact with iceberg	4	23.6
World's fastest elevator	5	37.6
Secretariat, world's fastest Triple Crown winner	6	39.7
Cheetah	7	60
Maximum legal speed on a U.S. freeway (may vary from state to state)	8	65
Hurricane force wind	9	73
World's fastest roller coaster (Kingda Ka Rocket Coaster at Six Flags Great Adventure, NJ)	10	128
Speeding bullet	11	200
Fastest official lap speed at Indy 500 (by Arie Luyendyk in 1996)	12	237
Commercial airline at cruising altitude	13	500
Speed of sound (but varies with atmospheric temperature)	14	760
F/A-16 fighter jet	15	1260

Activities Featuring Algebra

A Game of Functions (1974)

by Robert Froman

Crowell, ISBN #069000544X

Temperature (1995)

by Brenda Walpole

Gareth Stevens, ISBN #0836813634

How Little and How Much: A Book about Scales (1976)

by Franklyn M. Branley

Crowell, ISBN #0690010583

Overview of Books: In *A Game of Functions,* discover the meaning of functions and how we encounter them every day in our world. Next, learn about the history of thermometers, heating, cooling, insulation, and more in *Temperature*. This book also includes a variety of experiments dealing with temperature that can be carried out at school or at home. Then, although *How Little and How Much: A Book about Scales* is geared toward a younger audience, this book provides an excellent, concise history of Fahrenheit's and Celsius's method of calibrating thermometers, along with a comparison of the Fahrenheit and Celsius scales.

Mathematical Concepts and Skills: functions, dependent and independent variables, linear equations, slope, y-intercept, algebra, solving for one unknown, data collection, graphing, measurement, using metric and customary units

Science Concepts and Skills: measuring and recording temperature, Fahrenheit, Celsius, degrees, thermometer, scale

Overview of Activities: Students discuss what a function is and give real-world examples of functions. Students measure and record the temperature of an object heating or cooling, and discover the functional relationship between Celsius and Fahrenheit. Students also learn the history of Gabriel Fahrenheit's and Anders Celsius's methods for creating the scales on their thermometers. Students heat different amounts of the same substance and predict and discover whether a substance's boiling point is a function of the amount of the substance heated.

National Mathematics Standards (2000):

Students in grades 6–8 should "identify functions as linear or nonlinear and contrast their properties from tables, graphs, or equations" (p. 395). Students should "recognize and generate equivalent forms for simple algebraic expressions and solve linear equations" (p. 395). Students should also "explore relationships between symbolic expressions and graphs of lines, paying attention to the meaning of intercept and slope" (p. 395) (Algebra Standard). Students in grades 6–8 should "understand both metric and customary systems of measurement, understand relationships among units, and convert from one unit to another within the same system" (p. 399) (Measurement Standard).

National Science Standards (1996):

Students in grades 5–8 should "use appropriate tools and techniques to gather, analyze, and interpret data" and they should also "develop descriptions, explanations, predictions, and models using evidence" (p. 145) (Science as Inquiry, Content Standard A). Students in grades 5–8 should develop an understanding that "a substance has characteristic proprieties, such as density, a boiling point, and solubility, all of which are independent of the amount of the sample" (p. 154) (Physical Science, Content Standard B).

Materials:

substance that can be heated or cooled, sternos or a heating element, thermometers scaled with both Fahrenheit and Celsius degrees, graph paper, rulers

Description of Activities:

1. Introduce the concept of function by explaining how some things depend on others. For example, if you want to win a race, then you must run very fast. In other words, winning the race depends on how fast you run. Ask students to give examples of dependence and functions by thinking of functional relationships they encounter in their world (e.g., If I do my chores, then I will get paid my allowance. Thus, getting paid is a function of doing one's chores. If I stay up very late, then I will be tired in the morning. Thus, how tired you are depends on how much sleep you get).
2. Read excerpts from *A Game of Functions,* giving students more examples and explanations of functions and how graphs can be used to represent functions.
3. Ask students what they know about thermometers, who created them, how they work, and so on. Describe how in 1593, Galileo invented the first thermometer (which he called a thermoscope), using the expansion and contraction of air in a bulb to move water in an attached tube. The water inside the bulb expanded as the temperature outside increased and contracted as the temperature cooled. The book, *Temperature,* provides a good description of a thermoscope.

4. Refer to pages 20–30 in *How Little and How Much: A Book about Scales*, which provides an excellent and concise description of how Fahrenheit created the Fahrenheit scale, followed by a description of how the Swedish astronomer, Andres Celsius, created his Celsius (or Centigrade) scale.
5. Students work in pairs to measure and record in a table several (approximately five or six) readings of a substance heating or cooling. Students should record the temperature in both degrees Fahrenheit and Celsius.
6. Using graph paper and a ruler, students create a graph, plotting the temperature in degrees Celsius on the x-axis and the temperature in degrees Fahrenheit on the y-axis. Students then draw a line of best fit through as many points as possible. Students choose two points on the line and compute the slope and y-intercept of the line. Students describe the functional relationship between Celsius and Fahrenheit, which is given by the linear equation, $F = \left(\frac{9}{5}\right)C + 32$, where C = degrees Celsius and F = degrees Fahrenheit. Challenge students to reexpress this same linear equation, showing how C (degrees Celsius) depends on F (degrees Fahrenheit) (i.e., $C = \left(\frac{5}{9}\right)F - 17.78$).
7. Using these equations, students can verify at what Celsius temperature water freezes by plugging in 32 degrees for F (temperature at which water freezes in Fahrenheit degrees). Students can then look at their thermometers and notice that 32 degrees Fahrenheit equals 0 degrees Celsius. Similarly, using these equations, students can compute what the boiling point for water is in degrees Fahrenheit by plugging in 100 degrees for Celsius (temperature at which water boils in Celsius). Students should find that 212 degrees Fahrenheit corresponds to 100 degrees Celsius. Using these equations, challenge students to compute normal body temperature in degrees Celsius (37 degrees), knowing that normal body temperature is 98.6 degrees Fahrenheit.
8. Students work in small groups and heat different amounts of the same substance until their boiling points are reached. Students first predict whether the boiling point will be higher if a larger amount of the substance is heated. Students measure and record timed interval readings of the temperature until the substance boils. Students should discover that the boiling point is *independent* of the amount of the substance heated, although the time the substance takes to reach its boiling point is *dependent* on the amount of the substance heated. Thus, a substance's boiling point is *not* a function of the amount of the substance heated, but the time it takes for a substance to reach its boiling point *is* a function of the amount heated.

Assessment:

- Did students provide examples of one variable depending on another (i.e., a functional relationship)?
- Did students accurately measure and record temperature readings?
- Did students correctly graph degrees Celsius versus degrees Fahrenheit on their graph?
- Did students find the linear function that relates temperature in degrees Celsius to temperature in degrees Fahrenheit?
- Did students correctly solve for one unknown using the equations that relate Celsius and Fahrenheit?
- Did students discover that the boiling point is not a function (is independent) of the amount of the substance, but the time it takes for the substance to boil is a function of (depends on) the amount of the substance heated?
- Did students take accurate readings of temperature?

Activity Extensions:

- Students allow a heated substance to cool, record its temperature at various timed intervals, and plot temperature (y-axis) versus time (x-axis) on graph paper. Students will find that objects cool not linearly (at a steady rate) but exponentially, a law discovered by Isaac Newton. In other words, objects cool quickly until they reach a steady, level temperature (representative of an exponential function).
- Research and present findings on other weather instruments and scales, such as anemometers (used to measure wind speed), barometers (used to measure air pressure), Beaufort wind scale (measures wind intensity), Saffir-Simpson Hurricane Scale (measures a hurricane's intensity), Rankine scale (scale of absolute temperature), and so on, as well as the inventors of these instruments and scales.
- Explore extreme temperatures in the hottest and coldest parts of the world in *Hottest Coldest Highest Deepest* (Jenkins, 1998).
- Explore the climate-related phenomena and extreme temperatures in Alaska in *Arctic Nights Arctic Lights* (Miller, 2003).

Cross-Curricular Connections:

Social Studies

- Research and discuss demographics and geographical facts about very hot (desert) and very cold (tundra) places.

Visual Arts

- Explore artwork depicting weather and the seasons featured in *A Child's Book of Art: Great Pictures: First Words* (Micklethwait, 1993).

Related Literature

Branley, F. (1976). *How little and how much: A book about scales.* New York: Crowell.

Challoner, J. (1997). *Hot and cold.* Austin, TX: Steck-Vaughn.

Froman, R. (1974). *A game of functions.* New York: Crowell.

Goldsmith, M. (2001). *Galileo Galilei* (Scientists who made history series). Austin, TX: Raintree Steck-Vaughn.

Jenkins, S. (1998). *Hottest coldest highest deepest.* Boston: Houghton Mifflin.

Miller, D. (2003). *Arctic nights arctic lights.* New York: Walker.

Schwartz, D. (2001). *Q is for quark: A science alphabet.* Berkeley, CA: Ten Speed Press.

Scieszka, J., & Smith, L. (1995). *Math curse.* New York: Penguin Books.

Scieszka, J., & Smith, L. (2004). *Science verse.* New York: Penguin Books.

Walpole, B. (1995). *Temperature.* Chicago: Gareth Stevens.

Related Instructional Resources

Bender, L. (2005). *Invention* (Eyewitness books series). New York: Dorling Kindersley.

Bingham, C. (2004). *Invention* (Eye wonder books series). New York: Dorling Kindersley.

Brewer, D. (2004). *Inventions* (100 things you should know about series). New York: Barnes & Noble Books.

Egan, L. (1999). *Inventors and inventions.* New York: Scholastic.

Ellyard, D. (1996). *Weather* (Eyewitness discoveries series). New York: Dorling Kindersley.

Flack, J. (1989). *Inventing, inventions, and inventors: A teaching resourcebook.* Englewood, CO: Teacher Ideas Press.

Harper, C. (2001). *Imaginative inventions: The who, what, when, where, and why of roller skates, potato chips, marbles and pie (and more!).* Boston: Little, Brown.

LaFontaine, B. (1998). *Great inventors and inventions* (100 things you should know about series). London: Dover.

Micklethwait, L. (1993). *A child's book of art: Great pictures: first words.* New York: Dorling Kindersley.

Turvey, P. (1994). *Inventions: Inventors and ingenious ideas* (Timelines series). London: Franklin Watts.

VanCleave, J. (2004). *Scientists through the ages.* Hoboken, NJ: Wiley & Sons

Wood, R. (2003). *Great inventions* (Discoveries series). New York: Barnes & Noble Books.

Related Websites

Andres Celsius
http://www.answers.com/topic/celsius

Beaufort Wind Scale
http://www.stormfax.com/beaufort.htm

Gabriel Fahrenheit
http://www.answers.com/topic/fahrenheit

Galileo Galilei
http://www.answers.com/topic/galileo-thermometer

Saffir-Simpson Hurricane Scale
http://www.nhc.noaa.gov/aboutsshs.shtml

William John Macquorn Rankine
http://www.answers.com/topic/rankine-1

William Thomson Kelvin
http://www.answers.com/topic/kelvin-temperature-scale

Starry Messenger (1996)

by Peter Sis

Farrar Straus Giroux, ISBN #0374470278

Galileo Galilei (Scientists Who Made History Series) (2001)

by Mike Goldsmith

Raintree Steck-Vaughn, ISBN #0739844164

Overview of Books: Enjoy two biographies about the Italian scientist, mathematician, astronomer, philosopher, and physicist, Galileo Galilei, whose revolutionary discoveries impacted the world's view of the universe, despite great opposition during his lifetime. *Starry Messenger,* which is the title of Galileo's first book, has won many notable book honors and is rich with fascinating illustrations and detailed copies of Galileo's famous sketches.

Mathematical Concepts and Skills: functions, dependence, period, frequency, measurement, graphing, prediction

Science Concepts and Skills: scientific inquiry, motions and forces, history of science

Overview of Activities: Students explore and discuss the life and accomplishments of Galileo Galilei, who determined the mathematical law that relates the length of a pendulum to the time of its swing. Students work with pendulums, observe and record measurements, and discern that the rate of movement depends on the length of the string and not on the weight of the pendulum. Students work in small groups and research and present findings on one of Galileo's discoveries or inventions.

National Mathematics Standards (2000): Students in grades 6–8 should "use symbolic algebra to represent situations and to solve problems" and should "identify functions as linear or nonlinear, and contrast their properties from tables, graphs, or equations" (p. 395). Students should "model and solve contextualized problems using various representations, such as graphs, tables, and equations" (p. 395) (Algebra Standard). Students in grades 6–8 should "select and apply techniques and tools to accurately find length and angle measure to appropriate levels of precision" (p. 399) (Measurement Standard). Students should also "solve problems that arise in other contexts" (p. 402) (Problem Solving Standard).

National Science Standards (1996): Students in grades 5–8 should develop an understanding that "energy is a property of many substances and is associated with . . . mechanical motion and . . . is transferred in many ways" (p. 155) and that "an object that is not being subjected to a force will continue to move at a constant speed in a straight line" (p. 154) (Physical Science, Content Standard B). Students in grades 5–8 should also develop an understanding that "many individuals have contributed to the traditions of science" and that "studying some of these individuals provides further understanding of scientific inquiry" (p. 171). Students should also understand that "tracing the history of science can show how difficult it was for scientific innovators to break through the accepted ideas of their time to reach the conclusion that we currently take for granted" (p. 171) (History and Nature of Science, Content Standard G). Also, as a result of activities, students should "develop abilities to do scientific inquiry" and "develop understandings about scientific inquiry" (p. 143) (Science as Inquiry, Content Standard A).

Materials: string, heavy nuts or washers, hangers, stopwatch, *Starry Messenger* worksheet

Description of Activities:

1. Ask students to share what they know about the life and accomplishments of Galileo Galilei. Read *Starry Messenger,* followed by pages 4–5 in *Galileo Galilei* (Scientists Who Made History Series). Facilitate a discussion about Galileo, his accomplishments, and how he, despite great scrutiny and condemnation, "marked the beginning of a new method of thinking, a method based on observation and experiment—the method of modern science" (Goldsmith, 2001, p. 5).
2. Set the stage for the upcoming activity by describing how Galileo, when he was not quite twenty, began observing the swinging movement of a bronze chandelier in the Pisa Cathedral and used his pulse to time how long it took to make a complete swing (see page 11 in *Galileo Galilei*). Ask students if they have ever seen a pendulum and where, and how they think a pendulum works. Explain that they will be making a pendulum and determining whether the weight of the bob hanging on the pendulum or the length of the string attached to the bob, or both, affects its period (the time it takes for the pendulum to complete one full swing).
3. Place students in pairs. Students measure and cut two pieces of string, one measuring $\frac{1}{2}$ meter and the other measuring 1 meter long. Give students three heavy nuts (or three washers) and a hanger. Students make predictions as to whether weight, the length of the string, or both affect the period. Students share their predictions and articulate and defend their reasoning.

4. Using the 1 meter string, students tie a knot around one nut and tie the other end of the string to the hanger. Students suspend the hanger such that the pendulum (the string with the nut tied to it) can move freely. Using a stopwatch, students measure and record the pendulum's period. (To make accurate measurements of the period, let the pendulum swing ten full swings and then divide the time obtained by ten. Also, students should repeat their measurement three or more times, until an accurate measurement of period is achieved.)

5. Students repeat the above step, but add all three nuts to the 1 meter long string. Students record their measurements on the *Starry Messenger* worksheet.

6. Students repeat the above steps for the $\frac{1}{2}$ meter long string, with one nut suspended from it, and then for three nuts suspended from it. Students record their measurements on the *Starry Messenger* worksheet.

7. Students view their data and draw conclusions. (If students do not notice a pattern or trend, students might repeat the experiment using two nuts or a longer or shorter string.) Students should notice that weight does not affect the period, but that the length of the string does. More specifically, the period is a function of the length of the string. Further, the period of the pendulum with the 1 meter long string is approximately 2 seconds, while the period of the pendulum with the $\frac{1}{2}$ meter string is approximately 1.4 seconds. In other words, by doubling the length of the string (from $\frac{1}{2}$ meters to 1 meter), the period increased by a factor of 1.4 (it changed from 1.4 to 2, or by a factor of approximately 1.4). Remind students that 1.4 is close to the square root of 2 (or 1.414 . . .).

8. Introduce the formula for period discovered by Galileo: $T = 2\pi\sqrt{\frac{L}{g}}$ (where T = period, L = length of string, g = the acceleration due to gravity, or 9.8 m/s^2). That is, the period of a pendulum is proportional to the square root of its length. Students should notice that the period is affected by the length of the string because length, L, is a variable in the formula for period. On the other hand, weight is not a variable in the formula for period and thus, weight does not affect the period. In viewing the formula, discuss with students that the longer the length of the pendulum, the greater the period, and that the shorter the length of the pendulum, the shorter the period. Thus, the relationship between period and length is described as proportional.

9. Ask students to take their pendulum, pull it to one side, and watch it move. When does the pendulum move slowest? Fastest? What makes the pendulum move toward the center? What makes the pendulum oscillate back and forth? Facilitate a discussion describing how gravity pulls the bob toward the center, causing it to accelerate. (This is why g, the acceleration due to gravity, appears in the formula for period.) Then, the momentum built up by the acceleration due to gravity causes the pendulum to swing in the opposite direction. As the pendulum swings, there is a transfer of energy, between kinetic energy and potential energy.

10. Refer to page 11 in *Galileo Galilei* and finish the story about Galileo observing the pendulum swinging in the Pisa Cathedral. Using Galileo's discovery, a doctor friend, Santorio Santorio, developed a device known as a pulsilogium, which he used to measure a person's pulse. Read pages 14–15 in *Galileo Galilei*, which describe how Galileo continued to experiment and record data involving falling objects. His method of observation and experiment is now referred to as the "scientific approach to observation" (Goldsmith, 2001, p. 14).

11. Continue to read select passages from *Galileo Galilei*, sharing other facts about and accomplishments of Galileo, in particular, his view of the solar system and his trouble with the Roman Catholic Church due to his revolutionary views.

Assessment:

- Did students record accurate measurements for the period?
- Did students notice that weight does not affect the period, but that the length of the string does?
- Did students notice the proportional relationship between period and length?
- Did students articulate reasonable explanations for the movement of the pendulum?

Activity Extensions:

- Students experiment with changing the starting angle of the pendulum (how far you pull the string back) and whether this affects the period.
- Explore pendulum motion using one of the Interactive Pendulum websites listed below.
- Enjoy the poems "Clock-Watching" in *Lunch Money and Other Poems about School* (Shields, 1995) and "I Made a Perpetual Motion Machine" in *A Pizza the Size of the Sun* (Prelutsky, 1996).

Cross-Curricular Connections:

Social Studies

- View the timeline of Galileo in *Galileo Galilei* (pp. 44–45) and discuss his impact on such fields as science, astronomy, and philosophy.

Visual Arts

- Explore the art period known as surrealism by viewing Salvador Dali's "melting" clocks.

Related Literature

Campbell, J., & Collison, C. (2005). *G is for galaxy: An out of this world alphabet*. Chelsea, MI: Sleeping Bear Press.

Goldsmith, M. (2001). *Galileo Galilei* (Scientists who made history series). Austin, TX: Raintree Steck-Vaughn.

Lewis, J. (2005). *Galileo's universe*. Monroe, WI: The Creative Company.

Maestro, B. (1999). *The story of clocks and calendars*. New York: HarperCollins.

Pettenati, J. J. (2006). *Galileo's journal, 1609–1610*. Watertown, MA: Charlesbridge.

Prelutsky, J. (1996). *A pizza the size of the sun*. New York: Greenwillow Books.

Scieszka, J., & Smith, L. (1995). *Math curse*. New York: Penguin Books.

Scieszka, J., & Smith, L. (2004). *Science verse*. New York: Penguin Books.

Shields, C. (1998). *Lunch money and other poems about school*. New York: Puffin Books.

Sis, P. (1996). *Starry messenger*. New York: Farrar Straus Giroux.

Steele, P. (2005). *Galileo: The genius who faced the inquisition*. Washington, DC: National Geographic Society.

Wells, R. (2003). *How do you know what time it is?* Morton Grove, IL: Whitman.

Related Instructional Resources

Bender, L. (2005). *Invention*. New York: Dorling Kindersley.

Fredericks, A., & Asimov, I. (2001). *Science fair handbook: The complete guide for teachers and parents*. Tucson: Good Year Books.

Furniss, T. (1999). *Atlas of space exploration*. Milwaukee: Gareth Stevens.

Gardner, R. (1999). *Science projects about math*. Springfield, NJ: Enslow.

Gardner, R. (2000). *Science projects about methods of measuring*. Berkeley Heights, NJ: Enslow.

Panchyk, R. (2005). *Galileo for kids: His life and ideas 25 activities*. Chicago: Chicago Review Press.

Schwartz, D. (2001). *Q is for quark: A science alphabet*. Berkeley, CA: Ten Speed Press.

VanCleave, J. (2004). *Scientists through the ages*. Hoboken, NJ: Wiley & Sons.

Van Rose, S. (2005). *Earth* (Eyewitness books series). New York: Dorling Kindersley.

Wood, R. (1995). *Great inventions*. New York: Barnes & Noble Books.

Related Websites

Galileo Galilei

http://galileo.rice.edu/sci/instruments/pendulum.html
http://www-history.mcs.st-andrews.ac.uk/Mathematicians/Galileo.html
http://scienceworld.wolfram.com/biography/Galileo.html

Interactive Pendulum

http://pbskids.org/zoom/games/pendulum/index.html
http://www.physics.uoguelph.ca/applets/Intro_physics/kisalev/java/pend1/index.html

Starry Messenger Worksheet

Name(s): ____

Using a stopwatch, record the period below for each trial.

Length of string (meters)	Number of Weights	Period
1	1	
1	3	
$\frac{1}{2}$	1	
$\frac{1}{2}$	3	

1. What effect does the weight of the bob have on the pendulum's period? Explain your answer.

2. What effect does the length of the string have on the pendulum's period? Explain your answer.

3. What happens to the period if you double the length of the string (from $\frac{1}{2}$ meter to 1 meter)? What conclusions can you draw?

Activities Featuring Geometry

Great Buildings (Discoveries Series) (2005)

by Anne Lynch

Barnes & Noble Books, ISBN #0760750726

Building Big (2000)

by David Macaulay

Houghton Mifflin, ISBN #0395963311

Overview of Books: Explore engineering facts and statistics about many great buildings and structures, from early empires to present day in *Great Buildings* (Discoveries Series) and *Building Big*.

Mathematical Concepts and Skills: drawing and building two- and three-dimensional objects, measurement, ratio, proportion, scale

Science Concepts and Skills: science and technology, contributions of technology

Overview of Activities: Students work collaboratively to design a scaled model or diorama of a great building or structure. Students present facts about their structure to the class and discuss the mathematics and technology used in the structure's engineering design.

National Mathematics Standards (2000): Students in grades 6–8 should "draw geometric objects with specified properties, such as side lengths or angle measures" and should "understand relationships among the angles, sides, lengths, perimeters, areas, and volumes of similar objects" (p. 397). Students should "recognize and apply geometric ideas and relationships in areas outside of the mathematics classroom, such as art, science, and everyday life" (p. 397) (Geometry Standard). Students in grades 6–8 should "solve problems involving scale factors, using ratio and proportion" (p. 399) (Measurement Standard).

National Science Standards (1996): Students in grades 5–8 should develop an understanding that "women and men of various social and ethnic backgrounds—and with diverse interests, talents, qualities, and motivations—engage in the activities of science, engineering, and related fields such as the health professions. Some scientists work in teams, and some work alone, but all communicate extensively with others" (p. 170). Further, students should develop an understanding that "scientists and engineers of high achievement are considered to be among the most valued contributors to their culture" (p. 171) (History and Nature of Science, Content Standard G).

Materials: posterboard, materials to make a scaled model or diorama (toothpicks, cardboard, clay, Styrofoam, fabric, foil, wax paper, glue, scissors, etc.)

Description of Activities:

1. Share some facts and images about some of the great buildings and structures in *Great Buildings* (e.g., the Mayan and Egyptian pyramids, the Parthenon, the Colosseum, the Great Wall of China, the Guggenheim Museum, various castles, cathedrals, palaces, skyscrapers, etc.) and in *Building Big* (e.g., Golden Gate Bridge, the Channel Tunnel, Hoover Dam, the Astrodome, etc.).
2. Students work in small groups to collaboratively research a particular building or structure that is considered an engineering marvel (e.g., bridge, tunnel, skyscraper, castle, monument, dam, dome, etc.). Students create a scaled model or diorama of their selected structure. Students should pay close attention to scale, as this is the key to making the display look true-to-life. Students articulate in a building journal how they scaled their structures.
3. Students present engineering facts about the structure (dimensions, capacity, function, etc.) and discuss the mathematics and technology used to create such a building. Students share their work with classmates.

Assessment:

- Did students create a scaled model or diorama of a great building or structure and articulately present their research findings to the class?

Activity Extensions:

- Explore a biography of a famous builder or architect such as Frank Lloyd Wright, Buckminster Fuller, etc., using one of the Famous Architects websites listed below.

Cross-Curricular Connections:

Social Studies

- Explore and discuss the need for laws and policies to govern scientific and technological applications, such as the safety and well-being of workers and consumers and the regulation of utilities.

Visual Arts

- Explore the use of the golden mean in the Acropolis and Parthenon and other ancient buildings and structures.

Related Literature

Bentley, D. (2005). *The seven wonders of the ancient world.* London: Oxford University Press.

Britton, T. (2002). *The World Trade Center* (Symbols, landmarks and monuments series). Edina, MN: ABDO.

Chrisp, P. (2006). *Pyramid* (DK experience series). New York: Dorling Kindersley.

Corrbishley, M. (1996). *The world of architectural wonders.* New York: Peter Bedrick Books.

Gerstein, M. (2003). *The man who walked between the towers.* Brookfield, CT: Roaring Brook Press.

Hyman, T. (2006). *Pyramids* (Wonders of the world series). Farmington Hills, MI: Kidhaven Press.

Laliberte, M. (2005). *Seven wonders of the ancient world.* Berkeley Heights, NJ: Enslow.

Levy, D. (2005). *The World Trade Center* (Great structures in history series). New York: Thomson Gale.

Lynch, A. (2005). *Great buildings* (Discoveries series). New York: Barnes & Noble Books.

Macaulay, D. (1975). *Pyramid.* Boston: Houghton Mifflin.

Macaulay, D. (1981). *Cathedral: The story of its construction.* Boston: Houghton Mifflin.

Macaulay, D. (1982). *Castle.* Boston: Houghton Mifflin.

Macaulay, D. (1983). *City: A story of Roman planning and construction.* Boston: Houghton Mifflin.

Macaulay, D. (2000). *Building big.* Boston: Houghton Mifflin.

Scieszka, J., & Smith, L. (1995). *Math curse.* New York: Penguin Books.

Scieszka, J., & Smith, L. (2004). *Science verse.* New York: Penguin Books.

Related Instructional Resources

Biesty, S. (2005). *Egypt in spectacular cross-section.* New York: Scholastic.

Hart, A., & Mantell, P. (1997). *Pyramids: 50 Hands-on activities to experience ancient Egypt.* Charlotte, VT: Williamson.

Salvadori, M. (2000). *The art of construction: Projects and principles for beginning engineers and architects.* Chicago: Chicago Review Press.

Saunders, H. (1988). *When are we ever gonna have to use this?* Palo Alto, CA: Dale Seymour.

Scholastic Books. (1995). *Architecture and construction: Building pyramids, log cabins, castles, igloos, bridges, and skyscrapers.* New York: Scholastic.

Schwartz, D. (2001). *Q is for quark: A science alphabet.* Berkeley, CA: Ten Speed Press.

Related Websites

Building Big

http://www.pbs.org/wgbh/buildingbig/index.html

Building a Diorama

http://www.ehow.com/how_12761_make-diorama.html
http://www.enchantedlearning.com/crafts/dwelling/

Famous Architects

http://www.architectstore.com/famous.html
http://architect.architecture.sk/

Khufu's Pyramid

http://www.pbs.org/wgbh/nova/pyramid/explore/khufuall.html
http://www.greatbuildings.com/buildings/Great_Pyramid.html
http://www.nationalgeographic.com/pyramids/khufu.html

Activities Featuring Measurement

"A Pizza the Size of the Sun"
(a poem in A Pizza the Size of the Sun) (1996)

by Jack Prelutsky

Greenwillow Books, ISBN #10590374699

Eclipse: Darkness in Daytime (1973)

by Franklyn Branley

Crowell, ISBN #0690254148

Overview of Poem and Book: "A Pizza the Size of the Sun" is the title of one of the many zany poems in Prelutsky's book of verse. Find out what a pizza the size of the sun might look like and just how big it is! Then, in *Eclipse: Darkness in Daytime,* learn the relationships between and among the earth, sun, and moon and how and why eclipses occur.

Mathematical Concepts and Skills: circumference, diameter, radius, area of a circle, ratio and proportion, fractions, decimals, percents, angle measurement

Science Concepts and Skills: solar eclipse; movement of and relationship between and among the earth, moon, and sun; corona; umbra; penumbra; Baily's Beads

Overview of Activities: Students create a pizza with several toppings and compute what fractional part (or percentage) of their pizza is covered with each topping. Students also discover the formula for the area of a circle. Students then model and provide explanations for solar eclipses.

National Mathematics Standards (2000): Students in grades 6–8 should "develop and use formulas to determine the circumference and the area of . . . circles" and "develop strategies to find the area of more-complex shapes" (p. 399). Students should "select and apply techniques and tools to accurately find . . . angle measures to appropriate levels of precision" (p. 399) (Measurement Standard). Students in grades 6–8 should "work flexibly with fractions, decimals, and percents to solve problems" (p. 393) (Number and Operations Standard).

National Science Standards (1996): Students in grades 5–8 should develop an understanding that "the sun is the major source of energy for phenomena on the earth's surface" (p. 161) and that "most objects in the solar system are in regular and predictable motion and that these motions explain such phenomena as the day, year, phases of the moon, and eclipses" (p. 160) (Earth and Space Science, Content Standard D).

Materials: compasses, protractors, rulers, calculators, scissors, penny, orange, Popsicle stick, lamp

Description of Activities:

1. Read the poem "A Pizza the Size of the Sun" to set the stage for the upcoming activities.
2. Although the sun is spherical in shape, when we think of a pizza, we are considering a circle. Students work in pairs to create the pizza in the poem, complete with all of the toppings mentioned, or their favorite pizza with a minimum of three toppings. Using a compass and a sheet of white paper, students sketch a large circle with a radius of 4 inches, serving as their pizza. Using a ruler, students draw radii, marking the location of each topping. (The toppings may overlap.)
3. Students estimate and record what fractional part (or percentage) of their pizza is covered with each topping. Using protractors, students measure and record the angle formed by each sector and then determine what fractional part of the whole pizza each sector represents. Remind students that there are 360 degrees in a circle. Thus, to compute the fractional portion that each sector represents, students divide the angle measure of the sector by 360 degrees. Students should check their calculations by verifying that all of the individual fractional sectors sum to one (or 100 percent). How close were their measurements to their estimations?
4. Students must now determine the *area* of each topping, assuming that their pizza is the size of the sun. Assist students in conceptually deriving the formula for area of a circle. (This activity assumes that students have discussed and understand the formula for the circumference of a circle.) Students draw a large circle on a new piece of paper and, using a ruler, slice it into eight equal sectors by drawing four diameters. Students cut the circle into its eight sectors. Tell students to arrange the eight sectors in a row such that alternating sectors point up (see Figure 1). Suggest to students that the shape of the circle now looks like a parallelogram. Explain to students that, if they were to cut the circle into additional sectors, say 16 or 32 sectors, the shape would even more closely resemble a parallelogram. Discuss how the area of

this parallelogram-like shape has the same area as the original circle since no sectors were removed or added.

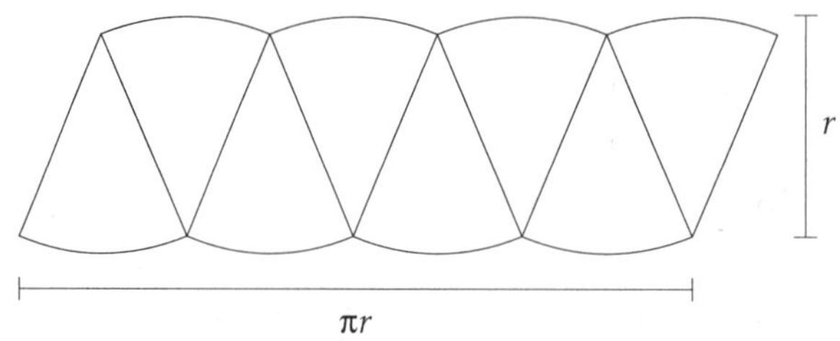

Figure 1. Rearrange the sectors to form a parallelogram.

5. Help students recall the definition of circumference: the distance around the outside of a circle, or $2\pi r$ (where r is the radius of the circle). Encourage students to see that since half of the sectors are pointed up (and half are pointed down), the base of the parallelogram is equal to one-half the circumference of the circle; that is, $\frac{1}{2}$ of $2\pi r$, or simply πr. The height of the parallelogram is the radius of the circle, or simply r. Since the area of a parallelogram is base times height, the area of this parallelogram-like shape is its base (πr) multiplied by its height (r), or πr^2. Thus, the area of a circle is πr^2.
6. Using the formula for area of a circle ($A = \pi r^2$), students compute the area of the entire pizza. Using the Internet (e.g., the NASA World Book—The Sun website listed below), students look up the radius of the sun (which is approximately 432,000 miles). Students substitute this value for the radius into the formula for the area of a circle and find that the area of a pizza the size of the sun is 5.86×10^{11} square miles. Students use this value to compute the area of each individual sector of the pizza (i.e., multiply the fractional amount each sector represents by 5.86×10^{11} to determine the area covered by that sector). Students should check their calculations by verifying that the sum of all of the individual sectors' areas equals the total area of the pizza; namely, 5.86×10^{11} square miles.
7. Begin a discussion about the sun. The sun, a star and the largest body in our solar system, is the major source of energy for phenomena on the earth including growth of plants, winds, ocean currents, and the water cycle.
8. Introduce the book *Eclipse: Darkness in Daytime*. Begin a discussion, asking students what a solar eclipse is and how it occurs. Read pages 1–5 in *Eclipse: Darkness in Daytime,* which briefly explain that, during a solar eclipse, the moon hides the sun. Challenge students to offer explanations as to how this can be since the moon is so much smaller than the sun.

9. Carry out an experiment similar to the one described on pages 6–7 in *Eclipse: Darkness in Daytime,* whereby students close one eye and look at a large object in the distance. Students hold a penny at arm's length in front of their eye and look at the object. Students should move the penny closer to their eye while continuing to focus on the large object in the distance. Students should notice that, as the penny is brought closer to their face, the object eventually is no longer visible. Relate this to the earth-moon-sun relationship in a solar eclipse. The moon (penny), despite it being so much smaller than the sun (large object in the distance), is so close to the earth (person holding the penny) that it appears to be the same size as the sun. Thus, during a total solar eclipse, the moon hides the sun.

10. Students model a solar eclipse by placing a light source (e.g., a lamp without its shade) in the center of the room, representing the sun. Students place an orange onto a Popsicle stick (or pencil), representing the moon. The students represent Earth. Students rotate the orange on the Popsicle stick around their head, modeling the moon revolving around Earth. When the moon is situated between the lamp (sun) and themselves (Earth), a solar eclipse occurs. Students should see that the sun is blocked, which results in darkness on Earth.

11. Using the Solar Eclipse Images and Video websites listed below, students view graphics of a solar eclipse and observe the changing appearance of the sun as the moon moves in front of it, blocking out its light. Define and discuss other solar eclipse terminology such as the corona (faint, white light that appears around the sun during a solar eclipse), the penumbra (the larger shadow cast by the moon during a solar eclipse), umbra (the darker, smaller shadow cast by the moon during a total solar eclipse), Baily's Beads (a chain of several bright "beads" of white light, immediately before and after a total solar eclipse), and so on.

12. Finish reading all or excerpts from *Eclipse: Darkness in Daytime,* which describes why ancient people feared eclipses and how to view a solar eclipse. Using the Solar Eclipse Calendar website listed below, find out when the next solar eclipse will occur and from where it can be seen.

Assessment:

- Did students estimate what fractional part (or percentage) of their pizza is covered with each topping?
- Did students derive the formula for area of a circle?
- Did students correctly compute the area of a pizza the size of the sun and the area of each of the sectors covered in toppings?
- Did students offer reasonable explanations for a solar eclipse?
- Did students correctly model a solar eclipse?

Activity Extensions:

- Challenge students to explain why solar eclipses do not occur more often. (This occurs because the earth's axis is tilted and, thus, the moon's shadow misses the surface of the earth most of the time.)
- Describe the differences between a total solar eclipse, an annular eclipse, and a partial eclipse.
- Model a lunar eclipse.
- Read *The Librarian Who Measured the Earth* (Lasky, 1994), which details how the Greek mathematician, Eratosthenes, approximated the circumference of the earth.

Cross-Curricular Connections:

Social Studies

- Research and discuss how ancient people interpreted and responded to eclipses.

Visual Arts

- Explore and imitate the heavenly skies in Vincent van Gogh's *A Starry Night.*
- Research and discuss why and how the sun is of prime consideration in architecture.

Related Literature

Branley, F. (1973). *Eclipse: Darkness in daytime.* New York: Crowell.

Kerrod, R. (2003). *Universe* (Eyewitness books series). New York: Dorling Kindersley.

Lasky, K. (1994). *The librarian who measured the earth.* Boston: Little, Brown.

Lippincott, K. (2004). *Astronomy* (Eyewitness books series). New York: Dorling Kindersley.

Prelutsky, J. (1996). *A pizza the size of the sun.* New York: Greenwillow Books.

Scieszka, J., & Smith, L. (1995). *Math curse.* New York: Penguin Books.

Scieszka, J., & Smith, L. (2004). *Science verse.* New York: Penguin Books.

Related Instructional Resources

Becklake, S. (1993). *The visual dictionary of the universe* (Eyewitness visual dictionaries series). New York: Dorling Kindersley.

Harrington, P. (1997). *Eclipse! The what, where, when, why and how guide to watching solar and lunar eclipses.* New York: Wiley.

Levine, S., & Johnstone, L. (2005). *First science experiments: Nature, senses, weather, and machines.* New York: Sterling.

Micklethwait, L. (1993). *A child's book of art: Great pictures: first words.* New York: Dorling Kindersley.

Schwartz, D. (2001). *Q is for quark: A science alphabet.* Berkeley, CA: Ten Speed Press.

Van Rose, S. (2005). *Earth* (Eyewitness books series). New York: Dorling Kindersley.

Related Websites

NASA World Book—The Sun
http://www.nasa.gov/worldbook/sun_worldbook.html

Online Copy of *A Pizza the Size of the Sun*
http://www.bookpage.com/9702bp/sponsor/apizzathesizeofthesun.html

Solar Eclipses
http://science.nasa.gov/newhome/headlines/ast05aug99%5F1.htm
http://www.exploratorium.edu/eclipse/index.html
http://www.kidseclipse.com/pages/a1b3c1d1.htm
http://www.kidseclipse.com/pages/a1b3c1d0.htm
http://sunearthday.nasa.gov/2006/educators/lp_68.php
http://solar-center.stanford.edu/eclipse/eclipse.html#activities

Solar Eclipse Calendar
http://sunearth.gsfc.nasa.gov/eclipse/eclipse.html

Solar Eclipse Images and Video
http://www.nasa.gov/vision/universe/solarsystem/sun_earthday2006.html
http://www.nasa.gov/vision/universe/solarsystem/sun_earthday2006.html

Sun Facts
http://www.kidseclipse.com/pages/a1b3c0d2.htm

Vincent van Gogh's *The Starry Night*
http://www.vangoghgallery.com/painting/starryindex.html
http://www.moma.org/collection/browse_results.php?object_id=79802

Human Body (DK Guide Series) (2001)

by Richard Walker

Dorling Kindersley, ISBN #0756622328

Overview of Book: Explore the human body via text, graphics, and close-up photographs in the fact-filled book, *Human Body* (DK Guide Series).

Mathematical Concepts and Skills: measurement, ratio, proportional reasoning, estimation, data collection

Science Concepts and Skills: structures and functions in living systems

Overview of Activities: Students explore, predict, and compute the ratio of their height to certain parts of their body to see how well their body ratios match those of Leonardo da Vinci's *Vitruvian Man*. Then students work collaboratively to research and present findings on a structure or function in the human body.

National Mathematics Standards (2000): Students in grades 6–8 should "understand, select, and use units of appropriate size and type to measure angles, perimeter, area, surface area, and volume" (p. 399). Students should "select and apply techniques and tools to accurately find length . . . to appropriate levels of precision" and "solve problems involving scale factors, using ratio and proportion" (p. 399) (Measurement Standard).

National Science Standards (1996): Students in grades 5–8 should develop an understanding that "the human organism has systems for digestion, respiration, reproduction, circulation, excretion, movement, control, coordination, and for protection from disease. These systems interact with one another" (pp. 156–157) (Life Science, Content Standard C).

Materials: measuring tapes, calculators, *Human Body* worksheet

Description of Activities:

1. Using one of the *Vitruvian Man* websites listed below, share a picture of Leonardo da Vinci's *Vitruvian Man*, his world-renowned drawing of a male inscribed in a circle and square, who exemplifies perfect proportions. Facilitate a discussion about how the height of the *Vitruvian Man* must equal his arm span since the *Vitruvian Man* is inscribed in a square. In other words, the ratio of height to arm span equals 1. Students stretch their arms out and view one another. Do they think the ratio of their height to arm span is 1:1?

2. Students work in pairs and measure and record their height on the *Human Body* worksheet. (Students need to convert their height from feet and inches into inches only.) Students complete the second column labeled "My Prediction" on the worksheet, where they make and record predictions about the measure of several of their body parts. Students then work in pairs to record the actual measurements (in the third column) of these various parts of their body. Students compute the ratio of their height to these various measured parts in the fourth column of the worksheet. Students may use calculators to assist them in computing ratios.
3. According to da Vinci, the *Vitruvian Man* possessed the following ratios:
   - Ratio of height to arm span (1:1)
   - Ratio of height to hand span (10:1)
   - Ratio of height to the distance from the top of the head to the bottom of the chin (8:1)
   - Ratio of height to the distance from the elbow to the armpit (8:1)
   - Ratio of height to the distance from the elbow to the tip of the hand (5:1)
   - Ratio of height to the maximum width of the shoulders (4:1)
4. Students discuss how accurate their predictions and computed measurements were, as well as their computed ratios as compared to the ratios of the *Vitruvian Man*. Which ratios matched? Which did not? Why might some of the *Vitruvian Man*'s ratios not match their body ratios exactly? How closely do their bodies match the proportions of the *Vitruvian Man*?
5. Read select pages from *Human Body*; in particular, page 4, which introduces the twelve systems in the human body (skin, skeletal, muscular, nervous, hormonal, blood, lymphatic, immune, respiratory, digestive, urinary, and reproductive).
6. Students work in pairs and select, research, and present facts using a booklet, diorama, slideshow presentation, or series of posters on a particular system of the human body.

Assessment:

- Did students reasonably estimate and accurately measure their body parts?
- Did students compute ratios correctly?
- Did students verify and articulate how well their body proportions matched those of the *Vitruvian Man*?
- Did students present accurate findings regarding their choice of a system in the human body?

Activity Extensions:

- The Greeks and Romans believed the "ideal man" was seven heads tall. Students measure their head (i.e., vertical length from the top of one's head to the chin), multiply this number by seven, and compare this product to their height. Are they "ideal"?
- Enjoy the poem, "One Inch Tall" (Silverstein, 2004). Then, students develop a short essay entitled, "If I Were Only One Inch Tall," complete with computations, in which they describe what one day would look like if they awoke and found they measured only one inch. Students describe how their world would need to change, in terms of the dimensions of objects, in order to live that day comfortably. Or, write an essay about a day in the life of an ant, beetle, or ladybug, through the insect's eyes.

Cross-Curricular Connections:

Social Studies

- Explore the timeline of major medical discoveries on page 62 in *Human Body*. Students explore and discuss a biography about one of the scientists named.

Visual Arts

- Explore artists' use of the golden mean in architecture (also known as the golden rectangle and the golden section).

Related Literature

Arnold, T. (2000). *Parts*. New York: Puffin Books.

Arnold, T. (2005). *More parts*. New York: Puffin Books.

Arnold, T. (2007). *Even more parts*. New York: Puffin Books.

McCallum, A. (2005). *Beanstalk: The measure of a giant*. Watertown, MA: Charlesbridge.

Schwartz, D. (2003). *Millions to measure*. New York: HarperCollins.

Scieszka, J., & Smith, L. (1995). *Math curse*. New York: Penguin Books.

Scieszka, J., & Smith, L. (2004). *Science verse*. New York: Penguin Books.

Silverstein, S. (2004). *Where the sidewalk ends*. New York: HarperCollins.

Related Instructional Resources

Bowen, D. (1991). *The visual dictionary of the human body* (Eyewitness visual dictionaries series). New York: Dorling Kindersley.

Casson, A. (2006). *Introduction to the human body*. Philadelphia: Chelsea House.

Mason, A. (1994). *Leonardo da Vinci: An introduction to the artist's life and work* (Barron's famous artist series). London: Aladdin Books.

Mason, A. (2004). *Leonardo da Vinci* (Lives of the artists series). Milwaukee: Gareth Stevens.

Rose, M. (2003). *Human body* (Discoveries series). New York: Barnes & Noble.

Schwartz, D. (2001). *Q is for quark: A science alphabet*. Berkeley, CA: Ten Speed Press.

Venezia, M. (1989). *Da Vinci* (Getting to know the world's greatest artists series). New York: Scholastic.

Walker, R. (2001). *Human body* (DK guide series). New York: Dorling Kindersley.

Related Websites

Golden Ratio

http://www.artlex.com/ArtLex/g/goldenmean.html
http://goldennumber.net/art.htm

Human Body

http://medtropolis.com/VBody.asp
http://www.innerbody.com/htm/body.html
http://www.bbc.co.uk/science/humanbody/

Vitruvian Man

http://www.bbc.co.uk/science/leonardo/gallery/vitruvian.shtml
http://leonardodavinci.stanford.edu/submissions/clabaugh/history/leonardo.html

Human Body Worksheet

Name(s): ____

My height is ____________ inches.

	My prediction (inches)	Measurement of body part (inches)	Ratio of height to body part (inches)
arm span			
hand span			
distance from top of head to bottom of chin			
distance from the elbow to the armpit			
distance from elbow to tip of hand			
maximum width of shoulders			

Which of your body ratios match those of the *Vitruvian Man*?

What might account for differences in your body ratios as compared to those of the *Vitruvian Man*?

Activities Featuring Data Analysis and Probability

Almost Gone: The World's Rarest Animals (2006)

by Steve Jenkins

Scholastic, ISBN #0439860571

Overview of Book: Discover which animals are already extinct and which are almost gone in this fascinating fact-filled book, which also offers explanations for several animals' extinction. At the end of the book, learn other facts about these extinct and nearly extinct animals, including their size and location of habitat.

Mathematical Concepts and Skills: estimation, data creation and interpretation, graphing, box and whiskers plot, average, mean, median, mode, range, extremes, lower quartile, upper quartile, outlier

Science Concepts and Skills: animal habitats, regulation and behavior of organisms, organisms' internal environment, survival and extinction of organisms

Overview of Activities: Students discuss extinct animals and those animals that are on the endangered species list, where these animals live, why they are extinct or endangered, and their animal classification. Students compute three different types of average (mean, median, and mode) and discuss which average is best, given a set of data. Students create and compare box plots, noting ranges, extremes in value, medians, and quartiles.

National Mathematics Standards (2000): Students in grades 6–8 should "formulate questions, design studies, and collect data about a characteristic shared by two populations or different characteristics within one population" (p. 401). They should "select, create, and use appropriate graphical representations of data including histograms, box plots, and scatter plots" (p. 401). Students should also "find, use, and interpret measures of center and spread, including mean and interquartile range" and "develop and evaluate inferences and predictions that are based on data" (p. 401) (Data Analysis and Probability Standard).

National Science Standards (1996): Students in grades 5–8 should develop an understanding of "regulation and behavior of organisms" and "diversity and adaptations of organisms" (p. 155). Students should understand that "regulation of an organism's internal environment involves sensing the internal environment and changing physiological activities to keep conditions within the range required to survive" (p. 157). Students should also understand that "an organism's behavior evolves through adaptation to its environment" and "how a species moves, obtains food, reproduces, and responds to danger are based in the species' evolutionary history" (p. 157) (Life Science, Content Standard C).

Materials: *Almost Gone* fact sheet, rulers, graphing calculators

Description of Activities:

1. Begin a discussion about what "extinction" and "endangered" mean. What animals are extinct and what animals are on the endangered species list? Encourage students to conjecture and offer reasons for why animal species become extinct or are nearly extinct. Pages 4–5 in *Almost Gone: The World's Rarest Animals* articulates the importance of how all living things are connected and offers reasons for why certain species have become, or are becoming, extinct.
2. Featured on each page in *Almost Gone: The World's Rarest Animals* is a different animal that is extinct or nearly extinct. After identifying the animal appearing on each page and showing its illustration to the class, challenge students to identify the animal's geographic home (e.g., Africa, Australia, etc.), the type of habitat in which the animal lives (e.g., grasslands, desert, marine, etc.), and its animal classification (e.g., mammal, reptile, etc.). Record this information on the *Almost Gone* fact sheet.
3. Read the brief passage that accompanies each animal, which provides a description of the animal's features and why these animals are almost gone, as the titles states. As you read, students record on the *Almost Gone* fact sheet the approximate number of each animal species that is still living. (Note: When recording this number, students should round; e.g., when Jenkins states, "fewer than 60 left," this should be approximated to 60 or some mutually agreed number close to 60.)
4. Students view the last column of data and estimate the average of these numbers (i.e., the average total number of endangered species still living). Ask students to share their estimations. How did they compute their average? Did they mentally compute a mean? Did they compute the median or mode?

5. Begin a discussion describing the three different types of averages: mean, median, and mode. Using the *Almost Gone* data, students compute each type of average. Which average do they think best represents the data? Students share their thinking and defend their reasoning. Ask students to describe how the piece of data for the Abington Island Tortoise affects each of the different types of average (because only one tortoise exists and this value could potentially be an outlier since it is so numerically far away from the other numbers). What would happen to each different type of average if this piece of data were deleted? What if the data for the bactrian camel were deleted?
6. Using the data, students create a box plot (see the Box Plots websites listed below for how to create and interpret a box plot). Students should identify the range, extremes, median, lower quartile, and upper quartile.
7. Students create a second box plot using just the mammal data and compare this box plot to the one created in the prior step. What values are the same? Different? Why have the range, extremes, quartiles, etc., changed? How do the medians compare? What conclusions can you draw when you compare the mammal data to the entire set of animal data?
8. Students select an animal on the endangered species list not featured in *Almost Gone* and collect facts about the animal (e.g., its habitat, reasons for near extinction, number remaining). Students share their findings with the class.

Assessment:

- Did students offer plausible reasons for animals' near extinction?
- Did students correctly identify each animal's classification, geographic home, and habitat?
- Did students correctly compute a mean, median, and mode as well correctly create and interpret a box plot?
- Did students correctly identify which type of average (mean, median, or mode) best represents the data?
- Did students accurately compare two box plots and make correct observations about the data?
- Did students articulately present their findings to the class about an animal on the endangered species list?

Activity Extensions:

- Using the size data at the end of Jenkins's book, students estimate and then compute the "average" (mean, median, and mode) size of the endangered animals featured in his book.
- Make a box plot of the nonmammals and compare and contrast it to the box plot of only mammals (or to the entire set of mammals).

Cross-Curricular Connections:

Social Studies

- Locate various animal habitats on a world map (e.g., tropical rain forests, deserts, grasslands, etc.).
- Research and discuss the implications of fast-growing populations on natural resources as well as animals' food and water supplies, etc.

Visual Arts

- Explore the works of wildlife artists, such as John James Audubon.

Related Literature

Jenkins, S. (1995). *Biggest, strongest, fastest.* Boston: Houghton Mifflin.

Jenkins, S. (2004). *Actual size.* Boston: Houghton Mifflin.

Jenkins, S. (2005). *Prehistoric actual size.* Boston: Houghton Mifflin.

Jenkins, S. (2006). *Almost gone: The world's rarest animals.* New York: Scholastic.

Jenkins, S., & Page, R. (2003). *What do you do with a tail like this?* Boston: Houghton Mifflin.

Kincade, S. (1992). *Our time is now* (Young people changing the world series). Upper Saddle River, NJ: Pearson Foundation.

Scieszka, J., & Smith, L. (1995). *Math curse.* New York: Penguin Books.

Scieszka, J., & Smith, L. (2004). *Science verse.* New York: Penguin Books.

Related Instructional Resources

Dorling Kindersley. (1998). *DK nature encyclopedia.* London: Dorling Kindersley Children.

Hare, T. (2005). *Animal fact file: Head-to-tail profiles of more than 90 mammals.* New York: Checkmark Books.

Lewis, B. (1998). *The kid's guide to social action: How to solve social problems you choose—and turn creative thinking into positive action.* Minneapolis: Free Spirit.

National Geographic Society. (2000). *National Geographic animal encyclopedia.* Hanover, PA: National Geographic Children's Books.

Schwartz, D. (2001). *Q is for quark: A science alphabet.* Berkeley, CA: Ten Speed Press.

Twist, C. (2005). *Reptiles and amphibians dictionary: An A to Z of cold-blooded creatures.* New York: Scholastic.

Related Websites

Animal Facts and Statistics

http://www.indianchild.com/amazing_facts.htm
http://lsb.syr.edu/projects/cyberzoo/a_list.html
http://www.factmonster.com/ipka/A0768508.html
http://sunsite.berkeley.edu/KidsClick!/midanim.html
http://www.kidsplanet.org/factsheets/map.html
http://www.sandiegozoo.org/animalbytes/index.html
http://www.worldalmanacforkids.com/explore/animals.html

Box Plots

http://nlvm.usu.edu/en/nav/frames_asid_200_g_3_t_5.html?open=instructions
http://regentsprep.org/regents/math/data/boxwhisk.htm
http://www.resample.com/xlminer/help/Boxplot/Box_Intro.htm

Create a Graph Online

http://nces.ed.gov/nceskids/createagraph/

Endangered Species

http://www.animalinfo.org/
http://www.sierraclub.org/wildlife/species/
http://www.fws.gov/endangered/kids/

Almost Gone Fact Sheet
Exploring Mean, Median, and Mode

Animal	Geographic Home	Habitat	Animal Classification	Approximate Number Left
Crested Shelduck				
Grand Cayman Blue Iguana				
Northern Hairy-Nosed Wombat				
California Condor				
Addax				
Yangtze River Dolphin				
Assam Rabbit				
Miami Blue Butterfly				
Javan Rhinoceros				
Golden Lion Tamarin				
Eastern Barred Bandicoot				
Giant Stick Insect				
Dwarf Water Buffalo				
Bactrian Camel				
Iriomote Cat				
Abington Island Tortoise				
Northern Right Whale				

Marie Curie (1994)

by Leonard Everett Fisher

Macmillan, ISBN #0027353753

Overview of Book: Discover the amazing life and accomplishments of Marie Curie, the Polish-French scientist, two-time Nobel Prize winner, and a discoverer of radioactive elements.

Mathematical Concepts and Skills: prediction, probability, randomness, Law of Large Numbers, data collection, graphing, data interpretation and analysis, halving, fractions, exponential decay, exponential functions

Science Concepts and Skills: properties and changes of properties in matter, transfer of energy, science and technology, science as a human endeavor, history of science

Overview of Activities: Students explore the life and accomplishments of Marie Curie and engage in a simulation where they explore probability, randomness, and exponential decay. Students also graph and discuss the nature of an exponential function and how it differs from a linear function. Students then research and present findings on scientists and nuclear physicists who were involved in the discovery and exploration of radioactive elements.

National Mathematics Standards (2000): π Students in grades 6–8 should "formulate questions, design studies, and collect data about a characteristic shared by two populations or different characteristics within one population" and "develop and evaluate inferences and predictions based on data" (p. 401). Students should also "use proportionality and a basic understanding of probability to make and test conjectures about the results of experiments and simulations" (p. 401) (Data Analysis and Probability Standard). Students in grades 6–8 should "work flexibly with fractions, decimals and percents, to solve problems" (p. 393) (Number and Operations Standard). Students in grades 6–8 should "identify functions as linear or nonlinear" (p. 395) (Algebra Standard).

National Science Standards (1996): Students in grades 5–8 should develop an understanding that "energy is a property of many substances and is associated with heat, light, electricity, mechanical motion, sound, nuclei, and the nature of a chemical. Energy is transferred in many ways" (p. 155). They should also develop an understanding that "in most chemical and nuclear reactions, energy is transferred into or out of a system. Heat, light, mechanical motion or electricity might all be involved in such transfers" (p. 155) (Physical Science, Content Standard B). Students in grades 5–8 should develop an understanding that "the effect of science on society is neither entirely beneficial nor entirely detrimental" and that "technological changes are often accompanied by social, political, and economic changes that can be beneficial or detrimental to individuals and society" (p. 169) (Science in Personal and Social Perspectives, Content Standard F). Students in grades 5–8 should develop an understanding that "women and men of various social and ethnic backgrounds—and with diverse interests, talents, qualities, and motivations—engage in the activities of science, engineering, and related fields such as the health professions. Some scientists work in teams, and some work alone, but all communicate extensively with others" (p. 170). Further, students should develop an understanding that "scientists and engineers of high achievement are considered to be among the most valued contributors to their culture" (p. 170) (History and Nature of Science, Content Standard G).

Materials: bags of fifty pennies, paper plates, cups, *Marie Curie* worksheet, graph paper

Description of Activities:

1. Ask students what they know about the life and accomplishments of Marie Curie. Read excerpts from the biography *Marie Curie*. Marie Curie was a French-Polish scientist who, along with the help of her chemist husband, Pierre, discovered the radioactive elements polonium (named after her native country of Poland) and the even more highly radioactive element, radium, in samples of pitchblende, the ore from which uranium is extracted. Marie Curie coined the term *radioactivity* after seeing a new kind of "ray" emanating from the brownish-black rock samples of pitchblende. Radioactivity is the spontaneous disintegration of the nucleus of an atom, which releases energy, mostly as radiation, which Marie Curie recorded.
2. Define the term *half-life,* which is the time it takes for half of a radioactive substance to decay. Different radioactive elements have half-lives of different lengths. Explain that if an element begins with 200 nuclei, after one half-life, half of the nuclei will have disintegrated, leaving 100. After the second half-life, half of the nuclei will have disintegrated, leaving 50, and so on.

3. Place students in pairs and give each pair a bag of fifty pennies (or M&M candies), representing 50 nuclei in a radioactive element. Using the *Marie Curie* worksheet, students first predict and record in the second column of the table how many nuclei they think will remain with each half-life until no nuclei remain (e.g., 25, then 12, then 6, etc.).
4. Students shake their bag of pennies and then pour them on a paper plate. If a penny lands heads-up (or if an M&M lands with the printed side up), the nuclei has decayed. Students discard these decayed nuclei, placing them aside in a cup. Students count the remaining pennies (or nuclei) and record this number on the *Marie Curie* worksheet in the third column. Students place these pennies (that have not decayed) back into the bag, shake the bag, pour onto the plate, and again remove any pennies that have decayed (are heads-up). Students record the number of remaining, nondecayed pennies. Students continue this experiment until no pennies are left, that is, until the radioactive element has totally decayed.
5. Students observe their table of values and discuss how long it took for their bag of pennies to decay. Did it take longer or shorter for their fifty nuclei to disintegrate as compared to their original prediction? Based on each pair's data, what was the average number of half-lives obtained by the class?
6. Students complete the right-hand side of the table by pooling the entire class's data. Students record the total number of nuclei used by the entire class and then predict and record how many nuclei they think will remain for each half-life (trial). Then, for each half-life, students sum the entire class's data and record this value in the last column of the table. Did the decay rate match their predictions? Did the decay rate change from their original experiment? Students should notice that the class's data (which represents a larger number of nuclei) better approximated the predicted behavior than did their individual trials using only fifty pennies. Discuss the Law of Large numbers (also called Bernoulli's Law), which states that if an experiment is carried out a large number of times, the empirical (or experimental) probability of the event occurring tends toward the theoretical probability of the event occurring.
7. Describe how the exact time it takes for a particular radioactive element to decay (its half-life) is the same, no matter how much or how little of the element you begin with. What *cannot* be predicted is which *particular* nuclei will decay. Also, a larger number of nuclei can better approximate the rate of decay.

8. Using graph paper, students graph the class's data (number of trials on the x-axis; number of remaining nuclei for each trial on the y-axis). Introduce this function as an example of an exponential function, one that decreases quickly but then levels off. Explain how this function differs from a linear function, one in which the rate of decay would be a constant, steady rate.

9. Students research and present a biography of an individual whose work centered on the study of radioactive elements or nuclear physics (e.g., Wilhelm Conrad Roentgen, Antoine Henri Becquerel, Pierre Curie, Ernest Rutherford, Albert Einstein, J. Robert Oppenheimer, Enrico Fermi, etc.). Or, students research and present findings on the advances in nuclear medicine due to the research and discovery of radioactive elements.

Assessment:

- Did students make reasonable predictions about the number of remaining nuclei?
- Did students correctly collect, record, and interpret the data?
- Did students understand why the Law of Large Numbers provides a closer approximation to the theoretical probability of an event occurring?
- Did students collect and present accurate findings about their research topic?

Activity Extensions:

- Students explore, collect data, and graph other examples of exponential functions, such as compound interest or the cooling of an object.
- Students research and explore the anatomy of an atom using the Atoms and Radioactive Decay websites listed below.
- Students research and present findings on other women who have made contributions to the fields of science, chemistry, physics, and medicine.

Cross-Curricular Connections:

Social Studies

- Explore and discuss the impact and repercussions of dropping the atomic bombs on Hiroshima and Nagasaki.
- Read and discuss *Snowflake Bentley* (Martin, 1998), a biography of Wilson Bentley, who, as a young boy, was fascinated with ice crystals and recorded their unique characteristics through microphotography over the course of his lifetime.

Visual Arts

- In art, color is the sensation produced by rays of light of different wavelengths. Explore the work of artists who have experimented with light and color (e.g., Picasso and his blue and rose periods, Claude Monet's paintings of the Rouen Cathedral, etc.).

Related Literature

Barasch, L. (2005). *Ask Albert Einstein*. New York: Frances Foster Books.

Brown, D. (1998). *Odd boy out: Young Albert Einstein*. Boston: Houghton Mifflin.

Coerr, E. (1997). *Sadako*. New York: Penguin Young Readers.

Coerr, E. (2004). *Sadako and the thousand paper cranes*. New York: Penguin Young Readers.

Fisher, L. (1994). *Marie Curie*. New York: Macmillan.

Graham, I. (2006). *Curie and the science of radioactivity*. New York: Barron's Educational Series.

Lassieur, A. (1998). *Marie Curie: A scientific pioneer* (Great life stories series). New York: Franklin Watts.

Martin, J. (1998). *Snowflake Bentley*. Boston: Houghton Mifflin.

Maruki, T. (1982). *Hiroshima no pika*. New York: HarperCollins.

Parker, S. (1995). *Marie Curie and radium* (Science discoveries series). New York: Chelsea House.

Scieszka, J., & Smith, L. (1995). *Math curse*. New York: Penguin Books.

Scieszka, J., & Smith, L. (2004). *Science verse*. New York: Penguin Books.

Steinke, A. (1987). *Marie Curie and the discovery of radium* (Solutions series). New York: Barron's Educational Series.

Tracy, K. (2004). *Pierre and Marie Curie and the discovery of radium* (Uncharted, unexplored and unexplained scientific advancements of the 19th century series). Hockessin, DE: Mitchell Lane.

Tsuchiya, Y. (1997). *Faithful elephants: A true story of animals, people and war*. Boston: Houghton Mifflin.

Related Instructional Resources

Burnie, D. (2000). *Light* (Eyewitness books series). New York: Dorling Kindersley.

Challoner, J. (2000). *Energy* (Eyewitness books series). New York: Dorling Kindersley.

Cooper, C. (1999). *Matter* (Eyewitness books series). New York: Dorling Kindersley.

Gardner, R. (2006). *Melting, freezing, and boiling science projects with matter*. Berkeley Heights, NJ: Enslow.

LaFontaine, B. (1998). *Great inventors and inventions* (100 things you should know about series). London: Dover.

Levine, S., & Johnstone, L. (2005). *First science experiments: Nature, senses, weather, & machines*. New York: Sterling.

Parker, S. (2000). *Medicine* (Eyewitness books series). New York: Dorling Kindersley.

Saunders, H. (1988). *When are we ever gonna have to use this?* Palo Alto, CA: Dale Seymour.

Schwartz, D. (2001). *Q is for quark: A science alphabet*. Berkeley, CA: Ten Speed Press.

VanCleave, J. (2004). *Scientists through the ages*. Hoboken, NJ: Wiley & Sons.

VanCleave, J. (2006). *Energy for every kid*. Hoboken, NJ: Wiley & Sons.

Related Websites

Atomic Bomb
http://inventors.about.com/library/weekly/aa050300a.htm

Atoms and Radioactive Decay
http://education.jlab.org/atomtour/

Historical Figures in Radioactivity
http://www.accessexcellence.org/AE/AEC/CC/historical_background.html

Interactive Radioactive Decay
http://www.colorado.edu/physics/2000/isotopes/radioactive_decay3.html

International Atomic Energy Agency
http://www.iaea.org/index.html

Marie Curie
http://nobelprize.org/nobel_prizes/physics/laureates/1903/marie-curie-bio.html

Nuclear Medicine
http://science.howstuffworks.com/nuclear-medicine6.htm

Pierre Curie
http://nobelprize.org/nobel_prizes/physics/laureates/1903/marie-curie-bio.html

Women in Science
http://womenshistory.about.com/library/bio/blbio_list_science.htm
http://www.astr.ua.edu/4000WS/4000WS.html
http://library.thinkquest.org/20117/mainbio.html

Marie Curie Worksheet

Name(s): ______________________________

Record your predictions for the number of remaining nuclei in the second column. Then, record the actual number of remaining nuclei after each trial in the third column.

Trial	Prediction	Number of Nuclei Remaining	Trial	Prediction	Number of Nuclei Remaining
Initial		50	Initial		
1			1		
2			2		
3			3		
4			4		
5			5		
6			6		
7			7		
8			8		
9			9		
10			10		
11			11		
12			12		
13			13		
14			14		
15			15		

1. Using the entire class's data, complete the right-hand side of the table, first recording your predictions for the number of remaining nuclei. Then, sum the class's data and record in the last column of the table the total amount of nuclei remaining after each trial.

2. What conclusions can you draw about computing the half-life of a radioactive element?

"I Made Something Strange with My Chemistry Set" (a poem in A Pizza the Size of the Sun) (1996)

by Jack Prelutsky

Greenwillow Books, ISBN #0590374699

So You Want to Be an Inventor? (2002)

by Judith St. George

Penguin, ISBN #0142404608

Perfect Inventions = Perfect World (2000)

by Fifth-Grade Students of Holly Springs Elementary School, Pickins, SC

Scholastic, ISBN #0439260663

Overview of Poem and Books: Laugh along with Prelutksy in his poem, "I Made Something Strange with My Chemistry Set," and learn about a disastrously funny chemistry experiment that backfired. Then, in *So You Want to Be an Inventor?*, discover what has been invented by both kids and adults, and how you could be the next inventor. In *Perfect Inventions = Perfect World*, the authors, who are middle school students and who view themselves as future leaders, scientists, and inventors, propose various inventions, machines, and technology that will protect and advance the planet while filling it with peace and happiness.

Mathematical Concepts and Skills: real-life applications of mathematics, careers in mathematics

Science Concepts and Skills: science and technology, inventions and inventors, science as inquiry

Overview of Activities: Students learn about and discuss a variety of inventions and how these discoveries and advances impact society today. Students then brainstorm and work collaboratively to design and illustrate a potential invention, with a focus on improving the quality of life and protecting the environment. Students identify the content knowledge and skills needed to make their invention a reality.

National Mathematics Standards (2000): Students in grades 5–8 should "recognize and apply mathematics in contexts outside of mathematics" (p. 402) (Connections Standard). Students should also "communicate their mathematical thinking coherently and clearly to peers, teachers, and others" and "use the language of mathematics to express mathematical ideas precisely" (p. 402) (Communications Standard).

National Science Standards (1996): Students in grades 5–8 should "develop abilities of technological design" and, in designing a solution or product, students should "make and compare different proposals in light of the criteria they have selected . . . consider constraints—such as cost, time, trade-offs, and materials needed—and communicate ideas with drawings and simple models" (p. 163). Students should also "evaluate completed technological designs or products" (p. 165) and "communicate the process of technological design" (p. 166) (Science and Technology, Content Standard E). Students in grades 5–8 should develop an understanding that "women and men of various social and ethnic backgrounds—and with diverse interests, talents, qualities, and motivations—engage in the activities of science, engineering, and related fields such as the health professions. Some scientists work in teams, and some work alone, but all communicate extensively with others" (p. 170). Further, students should develop an understanding that "scientists and engineers of high achievement are considered to be among the most valued contributors to their culture" (p. 171) (History and Nature of Science, Content Standard G).

Materials: Internet (or library resources), posterboard, crayons or markers

Description of Activities:

1. Read the poem "I Made Something Strange with My Chemistry Set" to set the stage for the upcoming activity in which students develop a potential invention. Briefly discuss obstacles individuals might encounter as they develop and put to the test new inventions and the importance of teamwork and perseverance.

2. Read excerpts from *So You Want to Be an Inventor?* After reading select excerpts about various inventors and their discoveries and inventions, engage students in a discussion about the implications of the inventions and their impact (positive or negative) on society.

3. Introduce the book *Perfect Inventions = Perfect World* to students, describing how the book was written by fifth graders who describe themselves on the opening page as "the next generation of leaders, scientists, and inventors." Remind students that they, too, are our future and can thus impact our planet, its environment, and its people, in both positive and negative ways. Engage students in a discussion of how they can impact our planet by asking such questions as, What are you doing now to protect the environment? Do you recycle? Do you conserve energy by not leaving on lights? Do you waste food? Water? Do you participate in volunteer organizations? Charitable organizations? After-school activities that foster good citizenship? Remind students that their actions today leave a lasting impact on others and on our planet; thus, they are in charge of making decisions for the common good.
4. Share with the class several of the clever and original ideas for inventions described and illustrated by the fifth-grade authors in *Perfect Inventions = Perfect World*. Note how the names of the students' inventions are derived from their Greek and Latin roots (which is why the names of the inventions may seem unfamiliar or awkward to pronounce). As you share various invention ideas, allow students to judge whether they think the invention is possible, beneficial, practical, and so on. Also, students should explain what content knowledge and skills they would need (e.g., in math, science, biology, medicine, botany, chemistry, etc.) in order to make the invention a reality.
5. Students work in small groups and brainstorm ideas of potential inventions. Each group selects one invention and, in the spirit of *Perfect Inventions = Perfect World*, names the invention using its Greek or Latin roots, describes it, and illustrates it. Students then present sketches of their inventions to the class and describe what mathematics and science skills they might need in order to make their invention a reality. To help direct students, consider narrowing the focus to be an invention that would help their school, neighborhood, state, or similar groups.

Assessment:

- Did students develop a potential original invention, complete with a name, description, and illustration or model?
- Did students develop a potential original invention that would benefit society?
- Did students clearly articulate the mathematical and scientific knowledge and skills needed to create their potential original invention?

Activity Extensions:

- Visit the U.S. Patent and Trademark Office website and research the patent process, how many patents are in existence, who has received the greatest number of patents, the first women to receive a patent, and similar questions. Also, research the difference between a patent, trademark, and copyright.
- Research and present the findings of women inventors using the literature resources below.
- Discuss students' choices of future careers. Focus on the mathematics and science they will need to know to perform well in these jobs. Also, discuss how these jobs contribute to the betterment of society or the environment.

Cross-Curricular Connections:

Social Studies

- Explore and discuss the inventions of and discoveries made by Archimedes.

Visual Arts

- Research and discuss the sketches of inventions of Leonardo da Vinci.
- Create a collage of individuals who have changed our world for the better through their inventions.

Related Literature

Barretta, G. (2006). *Now and Ben: The modern inventions of Ben Franklin?* New York: Holt.

Bender, L. (2005). *Invention* (Eyewitness books series). New York: Dorling Kindersley.

Bingham, C. (2004). *Invention* (Eyewonder books series). New York: Dorling Kindersley.

DeMauro, L. (2002). *Thomas Edison: A brilliant inventor* (Time for kids series). New York: HarperCollins.

Fifth-Grade Students of Holly Springs Elementary School, Pickens, SC. (2000). *Perfect inventions = perfect world.* New York: Scholastic.

Garrison, J., & Tubesing, A. (1996). *A million visions of peace: Wisdom from the Friends of Old Turtle.* New York: Scholastic.

Harper, C. (2001). *Imaginative inventions: The who, what, when, where, and why of roller skates, potato chips, marbles and pie (and more!).* Boston: Little, Brown.

Krull, K. (1995). *Leonardo da Vinci.* New York: Penguin Books.

LaFontaine, B. (1998). *Great inventors and inventions* (100 things you should know about series). London: Dover.

Marzollo, J. (1994). *My first book of biographies: Great men and women every child should know.* New York: Scholastic.

Mason, A. (2004). *Leonardo da Vinci* (Lives of the artists series). Milwaukee: Gareth Stevens.

Prelutsky, J. (1996). *A pizza the size of the sun.* New York: Scholastic.

Reimer, L., & Reimer, W. (1995). *Mathematicians are people, too.* Palo Alto, CA: Dale Seymour.

Satterfield, K. (2005). *Benjamin Franklin: A man of many talents* (Time for kids series). New York: HarperCollins.

Schonberg, M. (2005). *I is for idea: An inventions alphabet.* Chelsea, MI: Sleeping Bear Press.

Scieszka, J., & Smith, L. (1995). *Math curse.* New York: Penguin Books.

Scieszka, J., & Smith, L. (2004). *Science verse.* New York: Penguin Books.

St. George, J. (2002). *So you want to be an inventor?* New York: Penguin.

Thimmesh, C. (2000). *Girls think of everything: Stories of ingenious inventions by women.* Boston: Houghton Mifflin.

Thimmesh, C. (2002). *The sky's the limit: Stories of discovery by women and girls.* Boston: Houghton Mifflin.

Turvey, P. (1994). *Inventions: Inventors and ingenious ideas* (Timelines series). London: Franklin Watts.

VanCleave, J. (2004). *Scientists through the ages.* Hoboken, NJ: Wiley & Sons

Wood, R. (2003). *Great inventions* (Discoveries series). New York: Barnes & Noble Books.

Wulffson, D. (2001). *Kid who invented the trampoline: More surprising stories about inventions.* New York: Penguin Young Readers Group.

Related Instructional Resources

Inventor and Inventions Series (Benchmark Books)

Anderson, M. (2006). *Amazing Leonardo da Vinci inventions you can build yourself.* White River Junction, VT: Nomad Press.

Brewer, D. (2004). *Inventions* (100 things you should know about series). New York: Barnes & Noble Books.

Casey, S. (2005). *Kids inventing! A handbook for young inventors.* San Francisco: Jossey-Bass.

Chipman, D. (1998). *Cool women.* Chicago: Girl Press.

Currie, S. (2001). *Women inventors.* San Diego, CA: Lucent Books.

Egan, L. (1999). *Inventors and inventions.* New York: Scholastic.

Flack, J. (1989). *Inventing, inventions, and inventors: A teaching resource book.* Englewood, CO: Teacher Ideas Press.

Kassinger, R. (2002). *Build a better mousetrap: Make classic inventions, discover your problem-solving genius, and take the inventor's challenge.* Hoboken, NJ: Wiley and Sons.

Levine, S., & Johnstone, L. (2005). *First science experiments: Nature, senses, weather, & machines.* New York: Sterling.

Lewis, B. (1995). *The kid's guide to service projects: Over 500 service ideas for young people who want to make a difference.* Minneapolis: Free Spirit.

Phillips, C., & Priwer, S. (2006). *The everything da Vinci book: Explore the life and times of the ultimate Renaissance man.* Avon, MA: Adams Media.

Saunders, H. (1988). *When are we ever gonna have to use this?* Palo Alto, CA: Dale Seymour.

Schwartz, D. (2001). *Q is for quark: A science alphabet.* Berkeley, CA: Ten Speed Press.

Welden, A. (1998). *Girls who rocked the world: Heroines form Sacagawea to Sheryl Swoopes.* Hillsboro, OR: Beyond Words.

Yount, L. (1996). *Twentieth century women scientists.* New York: Facts on File.

Related Websites

Biographies of Famous Mathematicians
http://www-groups.dcs.st-and.ac.uk/~history/BiogIndex.html

Biographies of Scientists
http://www.blupete.com/Literature/Biographies/Science/Scients.htm
http://www.infoplease.com/spot/scibio8.html
http://www.maths.tcd.ie/pub/HistMath/Links/SciBiog.html

Biographies of Women in Science
http://www.awm-math.org/biographies.html#sciencebios

Great Inventors
http://library.thinkquest.org/5847/homepage.htm
http://library.christchurch.org.nz/Kids/Inventions/WhoInvented/
http://www.enchantedlearning.com/inventors/medicine.shtml

U.S. Patent and Trademark Office
http://www.uspto.gov/

Women Inventors
http://inventors.about.com/library/blwomeninventors.htm
http://inventors.about.com/od/womeninventors/index_a.htm
http://www.ideafinder.com/features/classact/women.htm

2

Literature-Based Mathematics and Social Studies Activities

The Mathematics-Social Studies Connection

Social studies encompasses many disciplines including anthropology, archaeology, economics, geography, history, law, philosophy, political science, psychology, religion, and sociology. Social studies also draws upon "appropriate content from the humanities, mathematics, and natural sciences" (NCSS, 1994, p. vii). According to the National Council for the Social Studies (1994), the primary purpose of teaching social studies is to enable learners to become informed consumers and skilled decision makers working toward the public good in a culturally diverse, democratic society coexisting in an interdependent world. Further, exemplary social studies programs prepare students to acquire, organize, interpret, and communicate information; process data and draw conclusions; and generate and assess alternative approaches to problem solving and making decisions (NCSS, 1994). These same skills of thinking, observing, hypothesizing, reasoning, communicating, and problem solving are skills the National Council of Teachers of Mathematics advocates in its *Principles and Standards for School Mathematics* (NCTM, 2000).

A growing body of research and anecdotal evidence documents the power of teaching social studies and mathematics in an integrated fashion, in particular, by using literature (Butzow & Butzow, 2006; Donoghue, 2001; Fredericks, 1991, 2000; Gallavan, 2001; Martinez & McGee, 2000; McCoy, 2003; Moss, 2003; Rose, 2000; Thompson & Holyoke, 2000; Ward, 2004c; Whitin & Whitin, 2004). Unquestionably, works of fiction and nonfiction offer a lens for students to see, understand, and experience their world. By integrating literature into your teaching, students can explore the many themes of social studies (such as culture, economics, geography, etc.) while learning the mathematics and statistics that support and define these themes.

"Children's literature enables students to explore the many themes of social studies while learning about mathematics and statistics."

This chapter articulates a variety of literature-based activities that integrate concepts and skills used and learned in the study of mathematics with those in social studies. While engaged in these activities, students will discover and gain practice with such mathematics concepts and skills as number sense, fractions, decimals, percents, ratios, proportions, exponents (Number and Operations Standard); pattern recognition, writing algebraic equations, variables, exponential and linear functions (Algebra Standard); two- and three-dimensional shapes, Pythagorean Theorem, angles, circles, diameter, altitude, slant height, symmetry, Cartesian coordinate system, coordinate graphing, x- and y-axis, ordered pairs (Geometry Standard); angle measurement, area, circumference, surface area, volume, scale (Measurement Standard); and data collection and interpretation, graphing, pie charts (Data Analysis and Probability Standard).

Social studies concepts and skills featured in this chapter include U.S. history; accomplishments and facts about U.S. presidents; purpose and characteristics of government; history and purpose of money and international currencies; international flags; world geography and statistics; map creation and interpretation; projection maps; population distributions; hunger and literacy rates; religions and languages in the world; inequities in resources and wealth and in access to resources; causes, repercussions, and implications of war; aspects and accomplishments of ancient civilizations; identification of scientific and technological advances; reconstructing the past using a variety of resources; and biography as a form of storytelling.

The integrated literature-based activities also provide students with many opportunities to predict, estimate, problem-solve, and reason (Problem Solving and Reasoning and Proof Standards) as well as communicate and use various representations to organize, record, model, and interpret mathematical ideas (Communication and Representations Standards). Further, students will discover and explore real-life applications of mathematics and social studies and careers in mathematics and social studies (Connections Standard).

Remember to check the appendix for ideas for assessment tools and rubrics.

Matrix of Mathematics and Social Studies Activities

BOOK TITLE	MATHEMATICAL CONCEPTS AND SKILLS	SOCIAL STUDIES CONCEPTS AND SKILLS	SCIENCE CONCEPTS AND SKILLS	VISUAL ARTS CONCEPTS AND SKILLS
Money (Eyewitness books series)	decimals, fractions, ratios, proportions	history and purpose of money and currency, various countries' currencies, geography	research and discuss the minting process and the process of creating paper money	explore, compare, and discuss the artwork of countries' currencies
One Grain of Rice	patterns, algebra, functions, exponential and linear functions, exponents, prediction, estimation	purpose of government, how government provides or does not provide for peoples' needs, characteristics of governments	research and discuss the implications of growing populations on resources and the environment	explore geometric patterns in Indian artwork
What's Your Angle, Pythagoras?	Pythagorean Theorem, writing an algebraic equation, variables, exponents	reconstructing the past using a variety of resources, biographies	explore and discuss Eratosthenes' method for approximating earth's circumference	research, discuss, and create a Pythagoras Tree fractal
Firefly Guide to Flags of the World	Cartesian coordinate system, coordinate graphing, *x*- and *y*-axis, origin, ordered pairs, symmetry	international flags, world geography and statistics, geographical features, map creation	explore and discuss the habitats and ecosystems of countries or regions	identify symmetry in international flags, explore and discuss the meaning of colors
Easy Origami; Sadako and the Thousand Paper Cranes	two- and three-dimensional shapes, spatial skills, angles (acute, right, obtuse, reflex, complementary, and supplementary), angle measurement, estimation	causes, repercussions, and implications of war	explore and discuss biographies of scientists involved in the development of the atomic bomb, research and discuss the ill effects of radiation	create a collage of images depicting war and its effects

BOOK TITLE	MATHEMATICAL CONCEPTS AND SKILLS	SOCIAL STUDIES CONCEPTS AND SKILLS	SCIENCE CONCEPTS AND SKILLS	VISUAL ARTS CONCEPTS AND SKILLS
The Seven Wonders of the Ancient World; Pyramid (DK Experience Series)	surface area, volume, pyramid, altitude, slant height, scale, ratio, proportion, number sense	aspects and accomplishments of ancient civilizations, Egyptian pyramids, Seven Wonders of the Ancient World, technological advances	explore and discuss the process of mummification, explore and discuss the tools used by builders of pyramids	explore Egyptian art, jewelry, and pottery; view and discuss works of artists who painted three-dimensional shapes
A World of Wonders: Geographic Travels in Verse and Rhyme; Maps and Mapping	measurement, area, perimeter, triangles, scale factors, ratio and proportion	map terminology (latitude, longitude, equator, scale, etc.), map reading, map creation, world geography	explore the scale of the solar system by computing distances between and among planets	view and discuss artists' renditions of maps
First Pets: Presidential Best Friends; Presidents (Eyewitness Books Series)	data interpretation and creation, pie charts, fractions, decimals, percents	U.S. history, accomplishments and facts about presidents who greatly influenced public policy	research and discuss presidents who were catalysts in improving environmental issues	view images of and locate facts about Mount Rushmore
If the World Were a Village	prediction, estimation, data interpretation, graphing, pie charts	geography, population distributions, hunger and literacy rates, religions and languages in the world, distribution of money, access to resources, inequities in resources and wealth	research and discuss which countries are impacted by malnutrition and famine and the implications; research and discuss the history of electricity and its role in advancing society	explore and discuss the aerial view artwork of Richard Diebenkorn; view, compare, and discuss artists' renditions of maps
The Sky's the Limit: Stories of Discovery by Women and Girls; Mathematicians Are People, Too; Five Brilliant Scientists (Great Black Heroes Series)	real-life applications of mathematics, careers in mathematics	identification of scientific and technological discoveries and advances, careers, biography as a form of storytelling	create a timeline of advances in medicine, science, or technology; research and discuss the inventions of Leonardo da Vinci	create a collage of careers that use knowledge of mathematics or science

Money (Eyewitness Books Series) (2005)

by Joe Cribb

Dorling Kindersley, ISBN #0756613892

Overview of Book: Through text and photographs, explore the diverse world of money, its history, currencies of various countries, and more in this fact-filled book.

Mathematical Concepts and Skills: decimals, fractions, ratios, proportions

Social Studies Concepts and Skills: history and purpose of money and currency, various countries' currencies, geography

Overview of Activities: Students explore and discuss the role and history of money and then create a travel brochure for an ideal vacation destination, complete with an itemized budget, images, demographic information, and the like. Students use ratios and proportions to convert between and among currencies.

National Mathematics Standards (2000): Students in grades 6–8 should "understand and use ratios and proportions to represent quantitative relationships" and "develop, analyze, and explain methods for solving problems involving proportions" (p. 393). Students should also "work flexibly with fractions, decimals, and percents to solve problems" (p. 393) (Number and Operations Standard).

National Social Studies Standards (1994): Social studies programs for the middle grades should also include experiences that provide for the study of *how people organize for the production, distribution, and consumption of goods and services,* so that the learner can "differentiate among various forms of exchange and money" and "describe the role of specialization and exchange in the economic process" (p. 96).

Materials: world map, calculators, Internet, or informational resources

Description of Activities:

1. Begin a discussion about money by asking such questions as, what is the role of money in society and in economies? Since paper money has little inherent worth, what gives money its value? In what forms does money come? What denominations are available in U.S. bank notes? Examine various bills and coins. What symbols, icons, and other text are featured on them? Why do you think these symbols, icons, and text were selected?

2. Begin reading the opening paragraph in *Money*, where students learn that money is a form of payment, whether it is a shell, coin, bill, or plastic card. Share some of the facts and pictures about the earliest money known to historians used in ancient Mesopotamia. Also, share select facts and pictures of other civilizations' forms of currency (such as iron hoes, cowrie shells, bars of rock salt, etc.) shown and described on pages 8–9.
3. Using the information presented on pages 10–13 in *Money*, share with students facts about and illustrations of the first coins and banknotes. Read other excerpts including the "Did you know?" facts (pages 64–66), the "Find out more" section (pages 68–69), and the "Timeline of banking" (pages 66–67).
4. Using a world map, students work in pairs and select a foreign country to which they would like to travel. They plan a one-week trip, along with an accompanying budget of expenses. Students research the cost for an airline ticket, hotel accommodations for six nights, food, transportation, entertainment (museums, sightseeing venues, etc.), passport fees, and other travel-related expenses. Students develop a two-column spreadsheet showing the cost of each item in U.S. dollars in the first column. Using one of the Currency Converter websites listed below, students locate and record the current exchange rate. Using ratios and proportions, students then compute the cost of each budgeted item in that country's currency, recording this figure in the second column of the spreadsheet. Students may use calculators to assist in their computations.
5. Students work in pairs to create a travel brochure (either a booklet or slideshow presentation) that includes their itemized budget, along with photos and other information about their country (e.g., geography, native language, population, cultural aspects, religion, industry, tourist attractions, etc.). Students should also include the name and an image of the country's currency, record its current exchange rate, and provide other historical and numerical facts about its currency.
6. Students present their travel brochures to the class. Students vote which vacation destination is most desirable, based on affordability, most interesting travel spot, and other features.

Assessment:

- Did students participate in a discussion about money?
- Did students perform correct calculations when pricing their travel costs, developing their budget, and converting between currencies?

- Did students work collaboratively to develop and articulately present their travel brochure?
- Did students present accurate facts about their travel destination?

Activity Extensions:

- Students examine the business page and collect, graph, interpret, and discuss data for a particular stock (or interest rates, etc.) over a period of time.
- Students design a new coin or banknote for a country of their choice, using symbols, text, and icons that best represent the country.
- Students write and then share an essay entitled, "A Day without Money."
- Students research a topic (e.g., the minting process, money in wartime, counterfeit money, checks and credit cards, shared currencies, etc.) and develop an artifact (e.g., poster, timeline, slideshow, etc.) to display and share their newly discovered information with the class.

Cross-Curricular Connections:

Science

- Research facts about and discuss the various metals and alloys used in the minting process.
- Research and discuss the process of creating paper money.

Visual Arts

- Explore, compare, and discuss the artwork of various countries' paper currencies.

Related Literature

Berg, A., & Bochner, A. (2002). *The totally awesome business book for kids*. New York: Newmarket Press.

Berg, A., & Bochner, A. (2002). *The totally awesome money book for kids*. New York: Newmarket Press.

Cribb, J. (2005). *Money* (Eyewitness books series). New York: Dorling Kindersley.

Harman, H. (2004). *Money sense for kids*. Hauppauge, NY: Barron's Educational Series.

Mayr, D. (2002). *Kids' everything money: From saving to spending to investing—Learn all about money!* Avon, MA: Adams Media.

Meltzer, M. (1993). *Gold: The true story of why people search for it, mine it, trade it, fight for it, mint it, display it, steal it, and kill for it.* New York: HarperCollins.

Scillian, D. (2003). *P is for passport.* Chelsea, MI: Sleeping Bear Press.

Related Instructional Resources

Bateman, K. (2001). *The young investor: Projects and activities for making your money grow*. Chicago: Chicago Review Press.

D'Alusio, F. (1998). *Women in the material world*. San Francisco: Sierra Club Books.

Menzel, P. (1995). *Material world: A global family portrait*. San Francisco: Sierra Club Books.

Related Websites

America Numismatic Association
http://www.money.org/AM/Template.cfm?Section=Home

Counterfeiting Protection
http://minneapolisfed.org/econed/curric/counter.cfm
http://www.moneyfactory.gov/newmoney/

Currency Converter
http://finance.yahoo.com/currency?u
http://www.oanda.com/convert/classic
http://www.gocurrency.com/currency-conversion.htm
http://www.xe.com/ucc/

FDIC for Students
http://www.fdic.gov/about/learn/learning/

Federal Reserve
http://www.federalreserveeducation.org/fed101/

Interactive European Shopping Adventure
http://www.econedlink.org/lessons/em299/flash/popupActivity.html

International Blank Note Society
http://www.ibns.it/

Money History, Minting, and Facts
http://www.pbs.org/wgbh/nova/moolah/
http://www.usmint.gov/kids/coinnews/mintingProcess/step_1.cfm
http://www.usmint.gov/kids/coinnews/theEuro.cfm
http://www.frbsf.org/currency/bills.html
http://library.thinkquest.org/28718/history.html
http://www.minneapolisfed.org/econed/curric/history.cfm
http://www.ex.ac.uk/~RDavies/arian/amser/chrono1.html

Activities Featuring Algebra

One Grain of Rice (1997)

by Demi

Scholastic, ISBN #059093998X

Overview of Book: Learn the tale of a wise and resourceful village girl who outsmarts a greedy raja by requesting for her reward one grain of rice for thirty days.

Mathematical Concepts and Skills: patterns, algebra, functions, exponential and linear functions, exponents, prediction, estimation

Social Studies Concepts and Skills: purpose of government, how government provides or does not provide for peoples' needs, characteristics of governments

Overview of Activities: Students predict the results of a doubling pattern and then express the doubling pattern as an exponential function. Students compare the growth of an exponential function to that of a linear function. Students also discuss, compare, and contrast various forms of governance and how these entities meet peoples' needs.

National Mathematics Standards (2000): Students in grades 6–8 should "represent, analyze, and generalize a variety of patterns with tables, graphs, words, and . . . symbolic rules" (p. 395). Students should "identify functions as linear or nonlinear and contrast their properties using tables, graphs, or equations" (p. 395). Students should "use symbolic algebra" and "model and solve contextualized problems using various representations such as graphs, tables, and equations" (p. 395) (Algebra Standard).

National Social Studies Standards (1994): Social studies programs for the middle grades should include experiences that provide for the study of *how people create and change structures of power, authority, and governance,* so that the learner can "analyze and explain ideas and governmental mechanisms to meet needs and wants of citizens, regulate territory, manage conflict, establish order and security" and to "explain and apply concepts such as power, role, status, justice, and influence to the examination of persistent issues and social problems" (p. 94).

Materials: *One Grain of Rice* worksheet, graphing calculators, map of world

Description of Activities:

1. Introduce *One Grain of Rice,* describing that it is a folktale about a girl from India. Show where India is located on a map. Discuss India's form of government, which is a Federal Republic. What are the characteristics of such a government?
2. Begin reading *One Grain of Rice*. Pause after reading that Rani asks for "two grains of rice, the next day four grains of rice, and so on for thirty days."
3. Record on the board students' predictions (without using any calculators) for the number of grains of rice Rani will receive on day 30. Ask students to explain and to justify their reasoning.
4. Students complete the *One Grain of Rice* worksheet where they record the number of grains of rice Rani receives each day. Students should notice the doubling pattern in the number of grains of rice received.
5. After completing the table, challenge students to work collaboratively to express the doubling pattern as an algebraic equation. (The doubling pattern can be expressed as $N = 2^{(d-1)}$, where N = number of grains of rice received and d = the number of the day). Define and explain such terminology as *function* and *dependence*. Explain how the number of grains of rice received by Rani is a *function* of the day; that is, the number of grains of rice Rani receives *depends on* what day it is in the sequence.
6. Using the equation, students compute the number of grains of rice Rani will receive on day 30 (a total of 536,870,912 grains). Students compare their original predictions to this number. How accurate were their predictions? Were students surprised at the enormity of this number and at how fast the doubling pattern grew? Explain that the equation that models the grains of rice Rani receives is an exponential function. Exponential functions grow slowly and then increase very quickly.
7. Graph the exponential function using a graphing calculator so that students can see the quickly increasing nature of this function. (Consider also graphing the linear equation $y = 2x$, so that students can compare the appearance of a linear function to an exponential function as well as the table of values associated with each of these functions.) Or, provide students with a copy of the final page in *One Grain of Rice,* which lists the number of grains of rice received each day. To demonstrate the quick growth of an exponential function, point out how on the tenth day, Rani received 512 grains of rice, and how on the twentieth day, she received over half a *million*. In other words, after only *ten* additional days, she received over one thousand times more grains of rice.

8. Challenge students to view a map of the world and to identify the forms of government that are in place in various parts of the world. What are the characteristics of such governments? (Consider viewing the World Factbook—Guide to Country Profiles website listed below for supplemental information.)
9. Compare and contrast aspects of a democratic society to one that is run by a tyrannical government. What are the advantages and disadvantages to both?
10. Students write a brief story describing a day in the life of someone living under tyrannical governance.

Assessment:

- Did students make reasonable predictions regarding the number of grains of rice Rani would receive and clearly explain their reasoning behind their predictions?
- Did students correctly compute the number of grains of rice received each day?
- Did students construct the correct exponential function that modeled the doubling pattern?
- Did students articulate the forms and characteristics of government in place in various parts of the world?
- Did students clearly articulate the advantages and disadvantages of various forms of governance?
- Did students paint an accurate portrayal of someone living under tyrannical rule?

Activity Extensions:

- At the end of *One Grain of Rice,* the author tells us that Rani received over one billion grains of rice. Students use their equation and a calculator to verify this. (Refer to the final page in *One Grain of Rice* for an explanation of the calculations.)
- Challenge students to compute the *cumulative* number of grains received with each successive day. Students look for a pattern and then express the pattern as an equation ($T = 2^d - 1$, where T = number of grains of rice received and d = the number of the day).
- Explore biographies of leaders (U.S. presidents, Adolph Hitler, Mussolini, Queen Elizabeth, Gandhi, etc.).
- Compare aspects of Indian culture to U.S. culture.

- Identify those countries that regularly face famine and discuss the implications. How can individuals or governments make a difference?
- Students write their own folktale.

Cross-Curricular Connections:

Science

- Research and discuss the implications of growing populations on resources and the environment.

Visual Arts

- Explore geometric patterns in Indian artwork.

Related Literature

Anno, M. (1999). *Anno's mysterious multiplying jar.* New York: Putnam Books.

Birch, D. (1988). *The king's chessboard.* New York: Puffin Books.

Demi. (1997). *One grain of rice.* New York: Scholastic.

Scillian, D. (2003). *P is for passport.* Chelsea, MI: Sleeping Bear Press.

Smith, D. (2002). *If the world were a village: A book about the world's people.* Tonawanda, NY: Kids Can Press.

Spier, P. (2002). *People.* New York: Doubleday.

Whelan, G. (2001). *Homeless bird.* New York: HarperCollins.

Related Instructional Resources

Chatterjee, M. (2002). *India* (Eyewitness books series). New York: Dorling Kindersley.

D'Alusio, F. (1998). *Women in the material world.* San Francisco: Sierra Club Books.

Ganeri, A. (2005). *India.* New York: Sea to Sea.

Menzel, P. (1995). *Material world: A global family portrait.* San Francisco: Sierra Club Books.

Related Websites

Doubling Pattern in *The King's Chessboard* (Birch, 1988)
http://www.beaconlearningcenter.com/Lessons/5107.htm

Find a Pattern with *One Grain of Rice* (Demi, 1997)
http://www.eduref.org/cgi-bin/printlessons.cgi/Virtual/Lessons/Mathematics/Patterns/PAT0200.html

USA Rice Federation—International Site Links
http://www.usarice.com/international.html

The World Factbook—Guide to Country Profiles
https://www.cia.gov/cia/publications/factbook/docs/profileguide.html

One Grain of Rice Worksheet

Name(s): ______________________________

Record the number of grains of rice Rani receives each day.

Day	Grains Received	Day	Grains Received
1	1	11	
2	2	12	
3		13	
4		14	
5		15	
6		16	
7		17	
8		18	
9		19	
10		20	

What is the pattern in the "Grains Received" column?

Express the pattern as a sentence or as an equation.

Using this sentence or equation, how many grains of rice will Rani receive on day 30?

What's Your Angle, Pythagoras? (2004)

by Julie Ellis

Charlesbridge, ISBN #1570911509

Overview of Book: In *What's Your Angle, Pythagoras?*, enjoy a story about a curious boy named Pythagoras whose chance encounter with a builder inspires him to develop the famous theorem named after him. This book includes an excellent geometric explanation of the Pythagorean Theorem.

Mathematical Concepts and Skills: Pythagorean Theorem, writing an algebraic equation, variables, exponents

Social Studies Concepts and Skills: reconstructing the past using a variety of resources, biographies

Overview of Activities: Students explore and discover a visual, geometric approach to deriving the algebraic equation known as the Pythagorean Theorem. Students use various resources to research and then report on biographies of famous mathematicians.

National Mathematics Standards (2000): Students in grades 6–8 should "represent, analyze, and generalize a variety of patterns with tables, graphs, words, and . . . symbolic rules" (p. 395). Students should "use symbolic algebra" and "model and solve contextualized problems using various representations such as graphs, tables, and equations" (p. 395) (Algebra Standard). Students in grades 6–8 should "create and critique inductive and deductive arguments concerning geometric ideas and relationships, such as . . . the Pythagorean relationship" (p. 397) (Geometry Standard).

National Social Studies Standards (1994): Social studies programs for the middle grades should include experiences that provide for the study of *the ways human beings view themselves in and over time*, so that the learner can "identify and use processes important to reconstructing and reinterpreting the past, such as using a variety of sources, providing, validating, and weighing evidence for claims, checking credibility of sources, and searching for causality" (p. 82). Social studies programs for the middle grades should include experiences that provide for the study of *relationships among science, technology, and society*, so that the learner can "examine and describe the influence of culture on scientific and technological choices and advancement, such as transportation, medicine, and warfare" (p. 99) and "show through specific examples how science and technology have changed people's perceptions of the social and natural world, such as in their relationship to the land, animal life, family life, and economic needs, wants, and security" (p. 99).

Materials: centimeter graph paper, scissors, rulers, calculators, *What's Your Angle, Pythagoras?* worksheet, map of the world, Internet or informational resources

Description of Activities:

1. Begin reading portions of *What's Your Angle, Pythagoras?* to set the stage for the hands-on exploration of the Pythagorean Theorem.
2. Place students in groups of threes. Distribute several sheets of centimeter paper to each group and ask students to cut and label squares with areas of 4, 9, 16, 25, 36, 49, 64, 81, 100, 121, 144, and 169 cm squared.
3. Challenge students to place three squares touching vertex to vertex, in an attempt to create a right triangle in the center. Once a group is able to create a right triangle using three squares, ask all students to build the same right triangle using the same three squares. Students record the areas of the three squares on the *What's Your Angle, Pythagoras?* worksheet. (The area of the squares serving as the legs of the right triangle should be recorded in the first two columns of the table and are called squares A and B. The area of the square serving as the hypotenuse of the right triangle should be recorded in the third column. This square is called square C.) Repeat this until students have discovered three or four combinations of squares that result in the creation of a right triangle.
4. Challenge students to look for a pattern in their table of values and to express it as an algebraic equation by determining the relationship between the three columns of data. Guide them in noticing that the square whose side serves as the triangle's hypotenuse has the largest area. Also, guide students in noticing that the sum of the areas of the squares (squares A and B) whose sides serve as the legs of the triangle equals the area of the square (square C) whose side serves as the hypotenuse (i.e., $A^2 + B^2 = C^2$, which is the Pythagorean Theorem). Challenge students to write this relationship as an equation using the variables A, B, and C, where A and B represent the lengths of the legs of the triangle and C represents the length of the hypotenuse of the triangle.
5. Share the illustrations in Ellis's book on pages 18–19 that support their findings.

6. Using one of the Pythagoras of Samos websites listed below, describe some of the other discoveries made by Pythagoras in the fields of mathematics, astronomy, and music theory. Then, using the Birthplaces of Famous Mathematicians website, locate Pythagoras' birthplace, the island of Samos, situated off the coast of Greece. By clicking on this location, find out what other mathematicians were born there (e.g., Aristarchus and Conon).
7. Break students into small groups and assign each group (or they select) a country or region of the world. Using the Birthplaces of Famous Mathematicians website as well as other informational resources, students write a biography (or create a poster or slideshow presentation) on a mathematician from that region, describing aspects of his or her life and his or her major accomplishments. Students present their findings to the class. Students might engage in role-play, where they dress up and act out a scene from their selected mathematician's life.

Assessment:

- Did students find combinations of three squares that formed a right triangle?
- Did students notice the pattern in the table of values?
- Did students accurately express the algebraic relationship between the three columns of data?
- Did students develop an algebraic and geometric understanding of the Pythagorean Theorem?
- Did students articulately present a biography on a mathematician?

Activity Extensions:

- Explore the Pythagorean Theorem websites below and discover other alternate ways to proving and visualizing the Pythagorean Theorem.
- Use *The Geometer's SketchPad* (if available) to dynamically explore the Pythagorean Theorem.
- Begin a unit examining ancient Greek civilization in terms of daily life, transportation, money and trade, architecture, culture, art, and similar aspects.

Cross-Curricular Connections:

Science

- Read and discuss *The Librarian Who Measured the Earth* (Lasky, 1994), which details how the Greek mathematician, Eratosthenes, approximated the circumference of the earth.

Visual Arts

- Research and discuss what a Pythagoras Tree is and how to create this fractal by visiting the Pythagoras Tree websites below.

Related Literature

Ellis, J. (2004). *What's your angle, Pythagoras?* Watertown, MA: Charlesbridge.

Lasky, K. (1994). *The librarian who measured the earth.* Boston: Little, Brown.

Reimer, L., & Reimer, W. (1995). *Mathematicians are people, too.* Palo Alto, CA: Dale Seymour.

Related Instructional Resources

Hart, A., & Mantell, P. (2004). *Ancient Greece! 40 hands-on activities to experience this wondrous age* (Kaleidoscope kids books series). Carmel, NY: Ideals.

Pearson, A. (2004). *Ancient Greece* (Eyewitness books series). New York: Dorling Kindersley.

Saunders, H. (1988). *When are we ever gonna have to use this?* Palo Alto, CA: Dale Seymour.

Schoefield, L. (2003). *Ancient Greece* (Discoveries series). New York: Dorling Kindersley.

Related Websites

Birthplaces of Famous Mathematicians
http://www-groups.dcs.st-and.ac.uk/~history/BirthplaceMaps/Places/Greece.html

Pythagoras of Samos
http://www-groups.dcs.st-and.ac.uk/~history/Mathematicians/Pythagoras.html
http://www.historyforkids.org/learn/greeks/science/math/pythagoras.htm

Pythagorean Theorem
http://www.pbs.org/wgbh/nova/proof/puzzle/theorem.html
http://www.utc.edu/Faculty/Christopher-Mawata/geom/geom7.htm
http://www.ies.co.jp/math/java/geo/pythagoras.html

Pythagoras Tree
http://www.ies.co.jp/math/java/geo/pytree/pytree.html

Virtual Manipulatives Library—Pythagorean Puzzles
http://nlvm.usu.edu/en/nav/frames_asid_164_g_3_t_3.html?open=instructions

What's Your Angle, Pythagoras? Worksheet

Name(s): ____

Area of Square *A* (whose side serves as a leg of the triangle)	Area of Square *B* (whose side serves as a leg of the triangle)	Area of Square *C* (whose side serves as the hypotenuse)

Do you notice a pattern in the table? (*Hint:* Look at columns 1 and 2 and compare them to column 3.)

Write an algebraic equation describing the relationship between the columns.

A triangle has legs measuring 15 units and 20 units and a hypotenuse measuring 25 units. Using the formula you just derived, prove this triangle is or is not a right triangle.

Activities Featuring Geometry

Firefly Guide to Flags of the World (2003)

by Firefly Books

Firefly Books, ISBN #1552978133

Overview of Books: Learn a plethora of facts and statistics about countries and states in this comprehensive guide, along with colorful maps and images of international and state flags.

Mathematical Concepts and Skills: Cartesian coordinate system, coordinate graphing, x- and y-axis, origin, ordered pairs, symmetry

Social Studies Concepts and Skills: international flags, world geography and statistics, geographical features, map creation

Overview of Activities: Students create and plot points on a Cartesian coordinate system. Students then create, sketch, and color a country's flag onto a Cartesian coordinate system, noting the ordered pairs associated with the boundaries of each colored region. Students are challenged to use another classmate's list of ordered pairs to recreate their classmates' flag. Students create a trifold poster, featuring geographic and demographic information about the country whose flag they graphed.

National Mathematics Standards (2000): Students in grades 6–8 should "use coordinate geometry to represent and examine the properties of geometric shapes" (p. 397) (Geometry Standard).

National Social Studies Standards (1994): Social studies programs for the middle grades should include experiences that provide for the study of *people, places, and environments* so that the learner can "create, interpret, use, and distinguish various representations of the earth, such as maps, globes, and photographs"; "use appropriate resources, data sources, and geographical tools . . . to generate, manipulate, and interpret information such as atlases, data bases, grid systems, charts, graphs, and maps"; "locate and describe varying landforms and geographic features, such as mountains, plateaus, islands, rain forests, deserts, and oceans, and explain their relationships within the ecosystem" (p. 85).

Materials: centimeter graph paper, rulers, color images of international flags, trifold posters, crayons, Internet or informational resources

Description of Activities:

1. Students begin by creating a rectangular coordinate system (also known as a Cartesian coordinate system). Distribute centimeter graph paper and rulers to students. Instruct students to place the paper landscape style on their desks and, using a ruler, draw a vertical line down the middle of the graph paper and then draw a horizontal line across the middle of the paper. Identify these lines as the *x*-axis (horizontal line) and the *y*-axis (vertical line). Students label the point where the lines intersect as (0, 0), also called the origin. Students create a scale by placing tick marks on the horizontal and vertical axes.
2. Describe how each point on the Cartesian coordinate system has a particular name that describes its location. This name is known as an ordered pair and is written as (x, y), where the x value tells you the distance on the *x*-axis (horizontal axis) from the origin, and the y value tells you the distance on the *y*-axis (vertical axis) from the origin.
3. Demonstrate how to plot a point on the Cartesian coordinate system by moving along the horizontal axis (to the left for a negative number or to the right for a positive number) and then along the vertical axis (down for a negative number or up for a positive number). Students practice plotting points.
4. Distribute to pairs of students (or let students choose) a color image of an international flag appearing in the *Firefly Guide to Flags of the World*. (Flags that work well are two- or three-colored striped flags or those without emblems; e.g., flags of Monaco or France.) Using a new piece of centimeter graph paper and rulers, students create a Cartesian coordinate system, sketch a flag of their choice, and color it in accordingly. Students label the corners of the various colored areas with their corresponding coordinate pairs. Students place the name of the flag at the top of the graph paper.
5. Students record all of the coordinate pairs on a separate sheet of paper (that match each colored region) and give this information to another pair of students, who are challenged to recreate and color the flag, using this list of coordinate pairs.
6. Students compare the flag they sketched and colored to the actual country's flag. How accurate were their flag recreations? How accurate were students in identifying the ordered pairs that denote the boundaries of each colored region?

7. Using *Firefly Guide to Flags of the World*, as well the Internet or other informational resources, students work in pairs to create a trifold poster, noting various geographic facts and statistics about the country whose flag they graphed. Students should sketch a map of the country, indicating the country's capital and its geographic features (e.g., mountains, plateaus, rain forests, deserts, etc.). Students should also note pertinent statistics (e.g., landmass area, capital, population, languages spoken, industry, economy, government, etc.) and research the choices of colors and symbols in their country's flag. What do the colors and symbols represent that appear on each country's flag? Students share their posters and findings with the class.

Assessment:

- Did students accurately create and scale a Cartesian coordinate system?
- Did students accurately record the ordered pairs that served as the boundaries for each colored region appearing on their flag?
- Did students accurately recreate a flag based on a list of ordered pairs?
- Did students present accurate information about their country?

Activity Extensions:

- Identify those countries whose flags have horizontal line symmetry (e.g., Sweden, Kuwait, Palestine, etc.), vertical line symmetry (e.g., Canada, Argentina, Suriname, etc.), or both (Japan, France, Israel, etc.). What flags are not symmetric (e.g., Bosnia & Herzegovina, United States, Solomon Islands, etc.)?
- Using *Material World: A Global Family Portrait* (Menzel, 1995), compare and contrast statistics of various countries included in this book. Facilitate a discussion addressing inequities in resources, health care, education, wealth, and like factors.
- Explore a biography of Rene Descartes, the French philosopher and creator of the Cartesian coordinate system.

Cross-Curricular Connections:

Science

- Explore and discuss the habitats and ecosystems of a particular country or region.

Visual Arts

- Students look for symmetry in international flags.
- Explore and discuss the meaning of various colors using the Meaning of Colors website listed below.

Related Literature

Crampton, W. (2002). *Smithsonian handbooks: Flags.* New York: Dorling Kindersley.

Crampton, W. (2003). *Spotter's guide to flags of the world.* London: Usborne.

Firefly Books. (2003). *Firefly guide to flags of the world.* Richmond Hills, Ontario: Firefly Books.

Glass, J. (1998). *The fly on the ceiling: A math myth.* New York: Random House.

Graham-Barber, L. (1992). *Doodle dandy! The complete book of independence day words.* New York: Bradbury Press.

Howe, R. (2002). *Flags of the fifty states and their incredible histories: The complete guide to America's most powerful symbols.* Guilford, CT: Lyons Press.

Smith, D. (2002). *If the world were a village: A book about the world's people.* Tonawanda, NY: Kids Can Press.

Smith, W. (2001). *Flag lore of all nations.* Brookfield, CT: Millbrook Press.

Westray, K. (2003). *Scholastic atlas of the world.* New York: Scholastic.

Znamierowski, A. (2001). *The world encyclopedia of flags.* London: Lorenz Books.

Related Instructional Resources

D'Alusio, F. (1998). *Women in the material world.* San Francisco: Sierra Club Books.

Haban, R. (1992). *How proudly they wave.* Minneapolis: Lerner.

Menzel, P. (1995). *Material world: A global family portrait.* San Francisco: Sierra Club Books.

Merriam Webster. (2003). *Webster's international atlas.* New York: Barnes & Noble.

Related Websites

International Flags
http://www.flags.net/index.htm

Meaning of Colors
http://www.princetonol.com/groups/iad/lessons/middle/color2.htm

Rene Descartes
http://www-groups.dcs.st-and.ac.uk/~history/Mathematicians/Descartes.html
http://www.renedescartes.com/

U.S. Flags
http://www.usflags.com/browsestore.asp?CategoryID=32

World Data and Statistics
http://www.infoplease.com/
http://www.worldbank.org/data/
http://www.cia.gov/cia/publications/factbook/
http://www.wri.org/
http://www.unicef.org/
http://www.un.org/
http://plasma.nationalgeographic.com/places/index.html
http://www.pbs.org/wgbh/nova/worldbalance/eart-nf.html
http://mapping.com/gv/
http://www.prb.org/
http://www.readtofeed.org/

Easy Origami (1992)

by John Montroll

Dover, ISBN #0486272982

Sadako and the Thousand Paper Cranes (2004)

by Eleanor Coerr

Penguin Young Readers, ISBN #0142401137

Overview of Books: Discover the ancient Japanese art of paper folding known as origami and create a variety of shapes, animals, and figures in *Easy Origami*. Then, learn the true story of a young girl named Sadako who lived in Hiroshima, Japan, when the atomic bomb was dropped during World War II. Upon being diagnosed with leukemia, Sadako set out on a quest to fold a thousand paper cranes, so that the gods would grant her wish to become well. This heartwarming story teaches young readers about the power of one person to create change, the long-term consequences of war, and values cherished by all cultures.

Mathematical Concepts and Skills: two- and three-dimensional shapes, spatial skills, angles (acute, right, obtuse, reflex, complementary, and supplementary), angle measurement, estimation

Social Studies Concepts and Skills: causes, repercussions, and implications of war

Overview of Activities: Students work in small groups to create an origami shape and estimate and measure its angles. Students also identify which angles are acute, right, obtuse, complementary, and supplementary. Students then learn of the true story of a young Japanese girl and her quest to make one thousand paper cranes in hopes to fight her leukemia, caused by the dropping of the atomic bomb on Hiroshima. Students discuss and debate whether the United States was justified in its use of the atomic bomb.

National Mathematics Standards (2000): Students in grades 6–8 should "precisely describe, classify, and understand relationships among types of two- and three-dimensional figures using their defining properties" and "understand relationships among the angles, side lengths, perimeters, areas, and volumes of similar figures" (p. 397). Students should also "recognize and apply geometric ideas and relationships in areas outside of the mathematics classroom, such as art, science, and everyday life" (p. 397) (Geometry Standard). Students in grades 6–8 should "select and apply techniques and tools to accurately find length and angle measures to appropriate levels of precision" (p. 399) (Measurement Standard).

National Social Studies Standards (1994): Social studies programs for the middle grades should include experiences that provide for the study of *how people create and change structures of power, authority, and governance*, so that the learner can "explain conditions, actions, and motivations that contribute to conflict and cooperation within and among nations" (p. 94).

Materials: origami paper, copy of folding instructions (appearing on page 6 in *Easy Origami*), protractors, map of Japan, two *New York Times* articles (see websites below)

Description of Activities:

1. Introduce students to origami, an ancient Japanese art of paper folding. Display on an overhead a copy (or make a hard copy for students) of the folding symbols used in origami appearing on page 6 in *Easy Origami*. Model each of the various folds with students.
2. Using the Origami Cranes website, students gain practice with the art of origami by creating an origami crane. (Students might also create the swan appearing on p. 30 in *Easy Origami*.)
3. Break students into small groups (two or three members) and give each group a copy of one page from *Easy Origami*, which displays the folding instructions for creating a particular figure. (Working in small groups will provide assurance that the students are folding their paper correctly as they create their origami figure.)
4. Students label each angle on their origami figure using letters of the alphabet. Students create a four-column table in their notebook where they list each angle by letter name (column 1), and then record their estimate of each angle's measure (column 2). Students use protractors to measure and record each angle (in column 3). (Column 4 is used in step 5.) As students perform each fold, they should identify any complementary or supplementary angles they encounter.

5. After creating their origami figure, students collaboratively compare their recorded angle measurements (in column 3), verifying the accuracy of their measurements. Students then compare their estimations for angle measurement (column 2) to the actual angle measurements (column 3). How accurate were their estimations for angle measure? In column 4 of their table, students classify each angles as acute, right, obtuse, or reflex.
6. Introduce the book *Sadako and the Thousand Paper Cranes,* informing students that this is a true story of a young girl named Sadako who lived in Hiroshima, Japan, in 1945 when the atomic bomb was dropped during World War II. Sadly, at age eleven, Sadako was diagnosed with leukemia, also known as the "atomic bomb disease." Inspired by a Japanese legend, Sadako begins folding a thousand paper cranes, trusting that the gods will grant her wish to become well and run again.
7. Using a world map, show the location of Japan and Hiroshima. Show the location of Nagasaki, the site of the second dropped atomic bomb. Facilitate a discussion as to why these sites were selected as targets.
8. Read all or excerpts from *Sadako and the Thousand Paper Cranes*. Pause to introduce and explain Japanese vocabulary or answer questions.
9. Using the *New York Times* websites listed below, distribute to students a copy of both articles that reported that the United States had dropped an atomic bomb on Hiroshima as well as Nagasaki. Facilitate a discussion or create a debate scenario in which students vocalize their thoughts on whether the United States was justified in its use of force and whether this type of force should be employed today.

Assessment:

- Did students accurately estimate and measure the angles in their origami figures?
- Did students participate in a discussion (or debate) about the use of the atomic bomb?

Activity Extensions:

- Students time how long it takes for them to create an origami crane. Students then compute how long it would take them to make 1,000 cranes, accounting for stopping for daily activities including, sleep, school, meals, and so on. Students share their calculations with the class. Are students surprised that Sadako completed 644 cranes, despite her illness?
- Students compose a haiku (three-line Japanese verse) describing a personal wish.
- Students create a timeline of the major events leading up to World War II.

Cross-Curricular Connections:

Science

- Explore and discuss a biography of a scientist involved in the development of the atomic bomb (e.g., J. Robert Oppenheimer, Arthur Compton, Enrico Fermi, etc.).
- Research the ill effects of radiation on humans, the earth, and the environment.

Visual Arts

- Create a collage of images depicting war and its effects.

Related Literature

Coerr, E. (1997). *Sadako*. New York: Penguin Young Readers.

Coerr, E. (2004). *Sadako and the thousand paper cranes*. New York: Penguin Young Readers.

Ishii, T. (2001). *One thousand paper cranes: The story of Sadako and the children's peace statue*. New York: Random House Children's Books.

Maruki, T. (1982). *Hiroshima no pika*. New York: HarperCollins.

Tsuchiya, Y. (1997). *Faithful elephants: A true story of animals, people and war*. Boston: Houghton Mifflin.

Related Instructional Resources

Bando, I. (1995). *Geometry and fractions with tangrams.* Vernon Hills, IL: Learning Resources.

Boursin, D. (2001). *Origami paper animals.* New York: Firefly Books.

Elfers, J., & Schuyt, M. (2000). *Tangrams: 1600 ancient Chinese puzzles.* New York: Barnes & Noble.

Ford, B. E. (1990). *The master revealed: A journey with tangrams.* Vallejo, CA: Tandora's Box Press.

Ford, B. E. (1990). *Tangrams: The magnificent seven piece puzzle.* Vallejo, CA: Tandora's Box Press.

Goodnow, J. (1994). *Math discoveries with tangrams.* Grand Rapids, MI: Ideal School Supply.

Kodansha International. (2002). *The first book of origami.* New York: Kodan-Sha America.

Mogard, S. (1992). *Windows to tangrams: Reproducible activities.* Covina, CA: American Teaching Aids.

Montroll, J. (1992). *Easy origami.* Mineola, NY: Dover.

Montroll, J. (1992). *Fun with easy origami: 32 projects and 24 sheets of origami paper.* Mineola, NY: Dover.

Press, J. (2001). *Around the world art and activities: Visiting the 7 continents through craft fun*. Charlotte, VT: Williamson.

Sarasas, C. (2002). *ABCs of origami.* North Clarendon, VT: Tuttle.

Yamauchi, Y. (1998). *Yami's origami: First steps to a thousand paper cranes*. Woodbridge, NJ: Woodbridge.

Related Websites

History of Origami
http://library.thinkquest.org/5402/history.html

New York Times Article (dated August 6, 1945)
http://www.nytimes.com/learning/general/onthisday/big/0806.html#article

New York Times Article (dated August 9, 1945)
http://www.nytimes.com/learning/general/onthisday/big/0809.html#article

Origami Cranes
http://www.savingcranes.org/teachers/kids/origami.cfm

Origami and Paper-folding.com
http://www.paperfolding.com/
http://dev.origami.com/diagram.cfm

Origami Space Telescope
http://www.californiaconnected.org/tv/archives/147

Activities Featuring Measurement

The Seven Wonders of the Ancient World (2002)

by Diana Bentley

Oxford University Press, ISBN #0195219147

Pyramid (DK Experience Series) (2006)

by Peter Chrisp

Dorling Kindersley, ISBN #0756614104

Overview of Books: In *The Seven Wonders of the Ancient World*, explore the largest and most impressive and advanced structures of early civilizations that have gained fame as the Seven Wonders of the Ancient World. Then, enjoy the panoramic sequences, digital imaging, and informative text as you travel down the Nile to the site of the remarkable Egyptian pyramids in *Pyramid* (DK Experience Series).

Mathematical Concepts and Skills: surface area, volume, pyramid, altitude, slant height, scale, ratio, proportion, number sense

Social Studies Concepts and Skills: aspects and accomplishments of ancient civilizations, Egyptian pyramids, Seven Wonders of the Ancient World, technological advances

Overview of Activities: Students explore facts about one of the Seven Wonders of the Ancient World, the Great Pyramids at Giza. Students make a net of a square pyramid and then calculate the surface area of Khufu's Pyramid. Students use ratios and proportions to relate the enormity of this pyramid to more familiar figures. Students work in small groups to research and present informational facts about one of the other Seven Wonders of the Ancient World.

National Mathematics Standards (2000): Students in grades 6–8 should "select and apply techniques and tools to accurately find length, area, and volume, and angle measures to appropriate levels of precision" and "develop strategies to determine the surface area and volume of selected prisms, pyramids, and cylinders" (p. 399) (Measurement Standard). Students in grades 6–8 should "recognize and apply mathematics in contexts outside of mathematics" (p. 402) (Connections Standard).

National Social Studies Standards (1994): Social studies programs for middle grades should include experiences that provide for the study of *the ways human beings view themselves in and over time,* so that the learner can "identify and describe selected historical periods and patterns of change within and across cultures, such as the rise of civilizations, the development of transportation systems, the growth and breakdown of colonial systems, and others" (p. 82). Social studies programs for the middle grades should also include experiences that provide for the study of *relationships among science, technology, and society,* so that the learner can "examine and describe the influence of culture on scientific and technological choices and advancement, such as transportation, medicine, and warfare" (p. 99).

Materials: paper, rulers, scissors, tape, calculators, Internet or informational resources

Description of Activities:

1. Challenge students to name the Seven Wonders of the Ancient World and to locate each of these wonders on a map. Why do students think these structures are considered "wonders"? Which ones are still standing? Share with students the names of as well as other brief facts about the Seven Wonders of the Ancient World appearing on pages 2–3 in *The Seven Wonders of the Ancient World.*

2. Read all or select text from pages 4–8 in *The Seven Wonders of the Ancient World,* which describes the Pyramids at Giza, the oldest and only remaining member of the Seven Wonders of the Ancient World, constructed over a twenty-year period, under the reign of the pharaoh, Khufu (also known by his Greek name, Cheops) around 2560 B.C. Show the location of Giza using a map. The Pyramids at Giza are a complex cluster of structures that include two temples in honor of Khufu, three smaller pyramids for Khufu's wives, an even smaller pyramid, a raised causeway connecting the two temples, and small mastaba tombs surrounding the pyramid for nobles. There was a town for the workers of Giza, including a cemetery, bakeries, a beer factory, and a copper smelting complex. More buildings and complexes are still being discovered by the Giza Mapping Project. Show students the image of Khufu's pyramid appearing on pages 12–13 in *Pyramid.*

3. Using paper, rulers, scissors, and tape, challenge students to make a net of a small-scale Egyptian pyramid. Discuss how an Egyptian pyramid is known as a *square* pyramid, which is a pyramid with four triangular sides and a square base. (A rectangular pyramid has a rectangle for its base and four triangles for sides, while a triangular pyramid, also known as a tetrahedron, has a triangular base and three triangular sides.)

4. Using their folded nets as a visual guide, challenge students to make a sketch of a square pyramid. Students label the pyramid with the measurements of Khufu's pyramid (height of 481 feet; length of one side of its base is 755 feet).

5. Tell students that archaeologists want to preserve the Khufu's pyramid by spraying it with a special sealant. Thus, archaeologists need to know how much sealant to buy in order to cover and protect the pyramid's entire surface. To solve this problem, students need to find the surface area of the pyramid. (Note: In computing the surface area of Khufu's pyramid, students will find the area only of the four triangular faces, but not the square base, since the square base is covered by ground.)

6. To find the surface area of the pyramid (less its square base), students must first calculate the *slant height* of one of its faces. Slant height is defined as the length of a line drawn perpendicular from the apex to one edge of the base. To compute the slant height, students must solve for the length of the hypotenuse using the Pythagorean Theorem, where one leg is the height of the pyramid (481 feet) and the other leg is one-half the distance of its base (377.5 feet). Thus, the slant height is approximately 611 feet. Knowing the slant height, students can calculate the surface area of the pyramid, which is the area of all four triangular faces (but not the square base in this case). The area of one triangular face is one-half times its base (755 feet) multiplied by its height (the slant height of 611 feet), which equals 230,652.5 square feet. So, the area of four triangular faces is four times this number, or 922,610 square feet. Thus, the surface area of Khufu's Pyramid (including just the four triangular faces, but not the square base) is 922,610 square feet. (If students want to find the surface area of the pyramid *including* its square base, then the area of the base is side times side, or 755 feet times 755 feet, which equals 570,025 square feet. Therefore, the square base plus all four triangular faces results in a total surface area of 1,492,635 square feet.) Thus, enough sealant would need to be purchased that would cover nearly one million square feet.

7. Using the statistics presented in *Pyramid,* challenge students to put the Great Pyramid's height, size, and surface area into perspective by relating its dimensions to more familiar objects. For example, Khufu's pyramid is nearly three times as tall as the Statue of Liberty. The average weight of a limestone pyramid block was 3.5 tons (approximate weight of two cars) and the pyramid is composed of approximately 2 million blocks. The lowest row of bricks measures (length, width, height) approximately 1.5 m × 1.5 m × 2.2 m (compared to the standard size of a brick, being 215 mm × 102.5 mm × 65 mm, or 8.5 in × 4 in × 2.6 in). The base of Khufu's Pyramid measures approximately 570,025 square feet. Since one acre is equivalent to 43,560 square feet, Khufu's pyramid encompassed over 13 acres (570,025 sq ft/43,560 sq ft), or approximately thirteen football fields.
8. Break students into small groups. Assign each group one of the other Seven Wonders of the Ancient World. Students create a poster, booklet, or slideshow showcasing pertinent facts about their Wonder, as well as images, and share their work and findings with the class.
9. Students vote on and discuss which of the Seven Wonders is the most amazing and why.

Assessment:

- Did students make a correct sketch of a net of a square pyramid?
- Did students correctly compute slant height and surface area of Khufu's pyramid?
- Did students place the immense size of the Great Pyramid into a more understandable perspective?
- Did students share accurate facts about their Wonder?

Activity Extensions:

- Present a scenario challenging students to compute the volume of Khufu's pyramid. The formula for the volume of a square pyramid is one-third times the area of the base (570,025 square feet) times the height (481 feet), or 91,394,008 cubic feet. Since the volume of an Olympic size pool is approximately 88,263 cubic feet, then over one thousand pools could fit inside Khufu's Pyramid!
- Challenge students to discover why a factor of one-third appears in the formula for the area of a pyramid.
- Students explore how the Nile River played a central role in the building of Khufu's Pyramid (refer to pages 22–29 in *Pyramid*).
- Students explore the mathematics and technology involved in building the other Seven Wonders.

Cross-Curricular Connections:

Science

- Explore and discuss the process of mummification (refer to pages 44–49 in *Pyramid*).
- Explore and discuss the tools used by builders of pyramids (refer to pages 20–21 in *Pyramid*).

Visual Arts

- Explore Egyptian art, jewelry, and pottery concealed in mummy's tombs (refer to pages 62–63 in *Pyramid*).
- Using *I Spy Shapes in Art* (Micklethwait, 2004), view and discuss works of artists who painted three-dimensional shapes (e.g., Andy Warhol's *Big Torn Campbell's Soup Can,* Joe Tilson's *Nine Elements,* Rene Magritte's *The Promenades of Euclid,* and M. C. Escher's *Hand with Reflecting Sphere*).

Related Literature

Bentley, D. (2002). *The seven wonders of the ancient world.* London: Oxford University Press.

Bingham, C. (2004). *Pyramid* (DK eye wonder series). New York: Dorling Kindersley.

Chrisp, P. (2006). *Pyramid* (DK experience series). New York: Dorling Kindersley.

Gibbons, G. (2004). *Mummies, pyramids, and pharaohs: A book about ancient Egypt.* New York: Little, Brown.

Hyman, T. (2006). *Pyramids* (Wonders of the world series). Farmington Hills, MI: Kidhaven Press.

Laliberte, M. (2005). *Seven wonders of the ancient world.* Berkeley Heights, NJ: Enslow.

Macaulay, D. (1975). *Pyramid.* Boston: Houghton Mifflin.

Morley, J., Bergin, M., & James, J. (1991). *An Egyptian pyramid.* New York: Bedrick.

Putnam, J. (2004). *Mummy* (Eyewitness books series). New York: Dorling Kindersley.

Putnam, J., Hayman, P., & Brightling, G. (2004). *Pyramid* (Eyewitness books series). New York: Dorling Kindersley.

Walker, J. (2003). *100 things you should know about ancient Egypt.* Broomall, PA: Mason Crest.

Related Instructional Resources

Biesty, S. (2005). *Egypt in spectacular cross-section.* New York: Scholastic.

Hart, A., & Mantell, P. (1997). *Pyramids: 50 Hands-on activities to experience ancient Egypt.* Charlotte, VT: Williamson.

Hart, G. (2003). *Ancient Egypt* (Discoveries series). New York: Barnes & Noble Books.

Hart, G. (2004). *Ancient Egypt* (Eyewitness books series). New York: Dorling Kindersley.

Honan, L. (1999). *Spend the day in ancient Egypt: Projects and activities that bring the past to life* (Spend the day in series). New York: Jossey-Bass.

Jordan S. (2000). *Ancient Egypt: Moments in history.* Logan, IA: Perfection Learning.

Micklethwait, L. (2004). *I spy shapes in art.* New York: Greenwillow Books.

Platt, R. (2003). *Hieroglyphics: The secrets of ancient Egyptian writing to unlock and discover* (Treasure chests series). Philadelphia: Running Press Kids.

Related Websites

Pyramid at Giza

http://www.pbs.org/wgbh/nova/pyramid/
http://www.greatbuildings.com/buildings/Great_Pyramid.html

Seven Wonders of the Ancient World

http://www.infoplease.com/ipa/A0001327.html
http://www.pbs.org/wgbh/nova/sunken/wonders/

A World of Wonders: Geographic Travels in Verse and Rhyme (2002)

by J. Patrick Lewis

Dial Books for Young Readers, ISBN #0803725795

Maps and Mapping (1985)

by Barbara Taylor

Kingfisher, ISBN #0753455064

Overview of Books: Enjoy a variety of poems about world geography, travelers, and map terminology in *A World of Wonders: Geographic Travels in Verse and Rhyme*. In *Maps and Mapping*, learn a brief history of the evolution of maps, how to read maps, and map terminology including equator, lines of latitude and longitude, contour lines, altitude, and sea level. This book also includes a collection of geography facts as well as experiments.

Mathematical Concepts and Skills: measurement, area, perimeter, triangles, scale factors, ratio and proportion

Social Studies Concepts and Skills: map terminology (latitude, longitude, equator, scale, etc.), map reading, map creation, world geography

Overview of Activities: Students explore features on a globe and engage in activities in which they read and interpret scale, compute the distances between cities and locations using a scale factor, and find the perimeter and area of the Bermuda Triangle and other landmasses. Students also explore and discuss the most popular type of map, a Mercator projection.

National Mathematics Standards (2000): Students in grades 6–8 should "understand, select, and use units of appropriate size and type to measure angles, perimeter, area, surface area, and volume" as well as "select and apply techniques and tools to accurately find length, area, volume, and angle measure to appropriative levels of precision" (p. 398). Students in grades 6–8 should also "solve problems involving scale factors, using ratio and proportion" (p. 398) (Measurement Standard).

National Social Studies Standards (1994): Social studies programs for the middle grades should include experiences that provide for the study of *people, places, and environments,* so that the learner can "create, interpret, use, and distinguish various representations of the earth, such as maps, globes, and photographs" and "use appropriate resources, data sources, and geographical tools such as aerial photographs, satellite images, geographic information systems (GIS), map projections, and cartography to generate, manipulate, and interpret information such as atlases, data bases, grid systems, charts, graphs, and maps" (p. 85).

Materials: globes, map of the Bermuda Triangle, string (or measuring tape), rulers, calculators, scaled Mercator projection of world, Internet or other informational resources

Description of Activities:

1. Set the stage for the activity by reading the poem "The Circle and the Poles" (or others) in *A World of Wonders: Geographic Travels in Verse and Rhyme.*
2. Place students in pairs and distribute globes. As described on pages 26–27 in *Maps and Mapping,* show the location and provide the definition of the lines of latitude, longitude, equator, the Great (or Prime) meridian, the Tropics of Cancer and Capricorn, and other features. Also, explore and discuss the symbols, legend, and scales on their globe. Share pages 8–11 in *Maps and Mapping,* which describe the importance of a map's scale and legend.
3. Read page 24 in *Maps and Mapping,* in which the author explains that a globe is the most accurate map of the world because both are round. Mapmakers (also called cartographers) experience difficulties when creating maps because they are trying to represent a three-dimensional figure (the earth) as a two-dimensional figure (a map). Thus, to create a map, they must cut, stretch, and distort some parts of the earth.
4. Students locate the following three places on their globe: the island of Bermuda, Miami (Florida), and San Juan (Puerto Rico), which designate the area known as the Bermuda Triangle. Students use string (or measuring tape) to measure the distances between and among each of the three locations. Students use the scale factor printed on the globe and compute the scaled distances between each location by setting up ratios and proportions. Students record the scaled distances between and among each of the three locations on a map of the Bermuda Triangle, using one of the Map of the Bermuda Triangle websites listed below.

5. Challenge students to compute the area encompassed by the Bermuda Triangle. Since the triangle is scalene, students will have to sketch and measure the altitude (height) of the triangle and then apply the formula for area of a triangle. Students might also compute the perimeter of the Bermuda Triangle.

6. Challenge students to compute the area and perimeter of other landmasses (e.g., Australia, United States, etc.) using their globes and string. Students check the accuracy of their measurements by comparing their calculated values to the actual landmass area using one of the World Atlas and Maps websites listed below. Students share their strategies for measuring and computing landmass, especially for those countries with jagged coastlines or boundaries.

7. Explain how Gerardus Mercator (born Gerhard Kramer), the German cartographer, introduced his Mercator projection of the earth in the 1500s, which became the standard world map for centuries. Mercator created his maps by projecting a globe, using light and shadows, onto a cylinder and then turning the cylinder into a flat map. (Consider cutting a cylindrical container vertically and opening it up for students to see that the result is a rectangle. Or, more simply, show how a rectangular piece of paper can be transformed into a cylinder by rolling it.) Describe the different types of projections (cylindrical, conic, and azimuthal) found on pages 28–29 in *Maps and Mapping*.

8. Students view a Mercator projection (see websites that follow activity) and make observations and discuss the distortions and inaccuracies in landmass. On a Mercator map, the lines of latitude and longitude run horizontally and vertically, respectively. However, a Mercator projection creates increasing distortions in size as one moves away from the equator and toward the poles. For example, encourage students to notice that on a Mercator map, Antarctica appears huge and Greenland appears significantly larger than South America, even though South America is eight times the size of Greenland; however, scale is true at the equator. Despite these inaccuracies, Mercator's maps were well received, in particular by sailors, since his maps preserve the shape of landmasses and directions, both of which are vital to ocean navigators.

Assessment:

- Did students correctly use their map's scale and proportions to compute the distances between cities and locations?
- Did students correctly compute the perimeter and area of triangles and the area of various landmasses?

- Did students clearly articulate their strategies for measuring landmass area?
- Did students notice the distortions in a Mercator projection?

Activity Extensions:

- Students measure and compute the circumference of the earth (approximately 24,900 miles) by wrapping a piece of string end-to-end around the earth and then using the globe's scale factor.
- As described on page 28 in *Maps and Mapping*, allow students to understand the struggles experienced by cartographers by peeling an orange (representing the earth) and making the peel lie flat (which is impossible). Students will find that they cannot flatten out the earth without breaking or dividing the peel into pieces.
- Students work in small groups to research and present findings on other types of map projections, including a Robinson projection (designed to provide an appealing depiction of the entire world, but has severe distortion at the poles which decreases as you move away from them), Peters projection (designed to represent landmasses of equal area equally, but not considered aesthetically pleasing), Mollweide projection (also known as the Babinet projection, or elliptical projection and is used when accurate representation of area takes precedence over shape), and the Winkel Tripel projection (current standard projection of the National Geographic Society and designed to show the earth with minimal distortion).
- Explore and report on a biography of a famous explorer (e.g., Columbus, Magellan, Marco Polo, etc.).
- Read other poems about maps, explorers, and map terminology in *A World of Wonders: Geographic Travels in Verse and Rhyme*.

Cross-Curricular Connections:

Science

- Discover how vast the universe is by viewing a map of the universe and computing distances between and among planets using the map's scale factor.

Visual Arts

- Visit the Six Contemporary Artists Who Use Maps in Their Work website and view and discuss several artists' renditions of maps (Alighiero Boetti, Kathy Prendergast, etc.).

Related Literature

Adler, D. (1991). *A picture book of Christopher Columbus.* New York: Scholastic.

Bramwell, M. (1998). *How maps are made* (Maps and mapmakers series). Minneapolis: Lerner.

Bramwell, M. (1998). *Mapping in everyday life* (Maps and mapmakers series). Minneapolis: Lerner.

Bramwell, M. (1998). *Mapping our world* (Maps and mapmakers series). Minneapolis: Lerner.

Carlisle, M. (1991). *Let's investigate marvelously meaningful maps.* New York: Barron's Educational Series.

Knowlton, J. (1985). *Maps and globes.* New York: HarperCollins.

Knowlton, J. (1988). *Geography from A to Z: A picture glossary.* New York: HarperCollins.

Kramer, S. (2004). *Who was Ferdinand Magellan?* (Who was . . . ? series). New York: Penguin Young Readers.

Lasky, K. (1994). *The librarian who measured the earth.* Boston: Little, Brown.

Lewis, J. (2002). *A world of wonders: Geographic travels in verse and rhyme.* New York: Dial Books for Young Readers.

Matthews, R. (2005). *Explorers* (Eyewitness books series). New York: Dorling Kindersley.

Ryan, P. (1989). *Explorers and mapmakers.* New York: Lodestar Books.

Sipiera, P. (1991). *Globes* (A new true book series). Chicago: Children's Press.

Taylor, B. (1993). *Maps and mapping* (Young discoveries series). New York: Kingfisher.

Worth, R. (2003). *The great empire of China and Marco Polo in world history.* Berkeley Heights, NJ: Enslow.

Related Instructional Resources

Ganeri, A. (1995). *Maps and mapmaking.* New York: Scholastic.

National Geographic Society. (2001). *Student atlas of the world.* Washington, DC: National Geographic Society.

Sammis, F. (2000). *Maps and mapmaking* (Mapping our world series). Tarrytown, NY: Marshall Cavendish.

Sobel, D. (1995). *Longitude: The true story of a lone genius who solved the greatest scientific problem of his time.* New York: Walker.

Related Websites

Map of the Bermuda Triangle
http://www.worldatlas.com/aatlas/infopage/printpage/bermudatriangle.htm
http://www.bermuda-triangle.org/assets/images/autogen/a_cart_1.gif
http://www.mysterieszone.com/pics/bermuda-triangle.jpg

Maps and Map Projections
http://www.colorado.edu/geography/gcraft/notes/mapproj/mapproj_f.html
http://atlas.nrcan.gc.ca/site/english/learningresources/carto_corner/index.html
http://www.enchantedlearning.com/geography/glossary/
http://www.btInternet.com/~se16/js/mapproj.htm
http://www.nationalgeographic.com/features/2000/exploration/projections/index.html
http://geography.about.com/library/weekly/aa031599.htm
http://www.flourish.org/upsidedownmap/
http://www.nationalgeographic.com/xpeditions/atlas/
http://www.kidsdomain.com/kids/links/Maps.html
http://worldatlas.com/aatlas/moremaps.htm
http://rockyweb.cr.usgs.gov/outreach/mapcatalog/
http://encyclopedia.kids.net.au/page/ma/Map_projection

Mercator Projection
http://science.nasa.gov/Realtime/rocket_sci/orbmech/mercator.html
http://worldatlas.com/aatlas/worldout.htm
http://www.warnercnr.colostate.edu/class_info/nr502/lg2/projection_descriptions/mercator.html

Mollweide Projection
http://worldatlas.com/aatlas/mollyout.htm
http://www.enchantedlearning.com/geography/glossary/

Outline Maps
http://www.eduplace.com/ss/maps/

Peters Projection
http://geography.about.com/library/weekly/aa030201a.htm
http://encyclopedia.kids.net.au/page/ga/Gall-Peters_projection

Printable Maps
http://www.nationalgeographic.com/xpeditions/atlas/

Robinson Projection
http://www.manifold.net/doc/7x/robinson.htm

Seven Continents
http://www.horizon-academy.com/Web_Quests/Around_The_World/html/the_continents.html

Six Contemporary Artists Who Use Maps in Their Work
http://www.artjunction.org/articles/mapartists.html

Winkel Tripel Projection
http://www.mapref.org/Winkel-Tripel-Projection.html

World Atlas and Maps
http://worldatlas.com
http://www.worldmapper.org/index.html
http://www.mongabay,com/igapo/world_statistics_by_area.htm

First Pets: Presidential Best Friends (2004)

by Nell Fuqua

Scholastic, ISBN #043959846X

Presidents (Eyewitness Books Series) (2000)

by James Barber

Dorling Kindersley, ISBN #0789488981

Overview of Books: In *First Pets: Presidential Best Friends,* learn the names and idiosyncrasies of the more than four hundred animals that have served as pets and faithful friends to U.S. presidents. Then enjoy Barber's fact-filled informative guide, *Presidents* (Eyewitness Books Series), which details the lives and accomplishments of the first forty-three men who have served as president of the United States.

Mathematical Concepts and Skills: data interpretation and creation, pie charts, fractions, decimals, percents

Social Studies Concepts and Skills: U.S. history, accomplishments and facts about presidents, presidents who greatly influenced and shaped public policy and decision making

Overview of Activities: Students create a pie chart depicting and comparing the pets owned by select U.S. presidents. Students then research facts about select presidents whom they believe to have made the most significant contributions during their terms in office regarding influencing and shaping public policy and decision making. A class debate ensues.

National Mathematics Standards (2000): Students in grades 6–8 should "formulate questions, design studies, and collect data about a characteristic shared by two populations or different characteristics within one population." They should "select, create, and use appropriate graphical representations of data including histograms, box plots, and scatter plots" and "develop and evaluate inferences and predictions based on data" (p. 401) (Data Analysis and Probability Standard).

National Social Studies Standards (1994): Social studies programs for the middle grades should include experiences that provide for the study of *the ideals, principles, and practices of citizenship in a democratic republic* so that the learner can "identify and explain the roles of formal and informal political actors in influencing and shaping public policy and decision-making" (p. 105).

Materials: paper, compass, crayons or markers, rulers, protractors, calculators, large posterboard

Description of Activities:

1. Ask students to make predictions about the various types of pets owned by U.S. presidents (e.g., most common pets, least common, number of pets owned per president, who owned the most pets, the least, etc.). Read from *First Pets* a few of the brief pet biographies and name some of the odd pets owned by presidents included in the book.
2. Copy select pages out of *First Pets* (e.g., pets owned by Jefferson, Lincoln, Hayes, T. Roosevelt, Coolidge) and place one president's pet page at a center. Students work in pairs at a center to create a pie chart displaying the pet data for that particular president. Students use a compass to draw a large circle on their paper (or prepare this in advance). In creating their pie charts, students will have to first sum the total number of pets owned by that president and then compute what fractional amount each pet (i.e., sector) represents. Students should convert each fraction to a decimal and then to a percent and color and label each sector of the pie chart with its corresponding percentage. Students may use calculators to assist them with their calculations.
3. After the pie charts are created, students share, compare, and make observations about their pie charts.
4. Share interesting facts about different presidents appearing in *Presidents*. Using the information as a springboard, students research and gather information about the major career highlights of a president. Students design a large poster to be used as a tool to convince classmates why their choice of president made the most significant contributions during his years in office in terms of influencing and shaping public policy and decision making. Facilitate a class debate during which students defend their choices for most influential president.

Assessment:

- Did students correctly display the pet data graphically as a pie chart?
- Did students make accurate observations when comparing pie charts?
- Did students present accurate findings about their president and defend why this president made the most significant contributions in terms of influencing and shaping public policy and decision making during his term?

Activity Extensions:

- Students create a timeline of the presidents and note important world events (wars, scientific discoveries, social issues, etc.).
- Students explore range and central tendency by creating a box plot of pets owned by a particular president.
- Using the InfoPlease Presidents website listed below, students create a line plot displaying presidential salaries. Is the president's salary growing linearly? Exponentially? Discuss how much the president should make, relative to individuals in other occupations and careers (e.g., teacher, professional athlete, astronaut, movie actor, medical doctor, etc.).
- Research and discuss the accomplishments of former first ladies.

Cross-Curricular Connections:

Science

- Research and discuss which presidents were catalysts in improving environmental issues.

Visual Arts

- View images of and locate facts about Mount Rushmore, located in South Dakota and created by the sculptor Gutzon Borglum, which features the heads of former presidents Washington, Jefferson, Theodore Roosevelt, and Lincoln.

Related Literature

Barber, J. (2004). *Presidents* (Eyewitness book series). New York: Dorling Kindersley.

Cheney, L. (2005). *A time for freedom: What happened when in America.* New York: Simon & Schuster.

Davis, G. (2004). *Wackiest White House pets.* New York: Scholastic.

Davis, K. (2002). *Don't know much about the presidents.* New York: HarperCollins.

Fuqua, N. (2004). *First pets: Presidential best friends.* New York: Scholastic.

McCullough, N. (2006). *Essential book of presidential trivia.* New York: Random House.

Reichblum, C. (2004). *Dr. Knowledge presents: Strange & fascinating facts about the presidents.* New York: Black Dog & Leventhal.

Rubel, D. (1994). *Scholastic encyclopedia of the presidents and their times.* New York: Scholastic.

Sobel, S. (1999). *How the U.S. government works.* Hauppauge, NY: Barron's Educational Series.

Sobel, S. (2001). *Presidential elections: And other cool facts.* Hauppauge, NY: Barron's Educational Series.

Sullivan, G. (1987). *Facts and fun about the presidents.* New York: Scholastic.

Thimmesh, C. (2004). *Madame President: The extraordinary, true, (and evolving) story of women in politics.* Boston: Houghton Mifflin.

Related Instructional Resources

King, D. (2007). *Have fun with the presidents: Activities, projects and fascinating facts.* Hoboken, NJ: Wiley & Sons.

Related Websites

Create a Graph Online
http://nces.ed.gov/nceskids/createagraph/

Mount Rushmore
http://www.mtrushmore.net/

Presidents
http://www.infoplease.com/ipa/A0873867.html
http://www.presidentsusa.net/
http://www.worldalmanacforkids.com/explore/presidents.html
http://www.whitehouse.gov/history/presidents/

Virtual Manipulatives Library—Pie Chart
http://nlvm.usu.edu/en/nav/frames_asid_183_g_2_t_5.html?open=activities

If the World Were a Village (2002)

by David J. Smith

Kids Can Press, ISBN #1550747797

Overview of Book: In Smith's book, which is subtitled, *A Book about the World's People*, the reader views the world as if it were a village of one hundred people. Along with the use of vividly colorful illustrations, Smith presents eye-opening statistics that dramatically portray the inequities in our world.

Mathematical Concepts and Skills: prediction, estimation, data interpretation, graphing, pie charts

Social Studies Concepts and Skills: geography, population distributions, hunger and literacy rates, religions and languages in the world, distribution of money, access to resources, inequities in resources and wealth

Overview of Activities: Students predict and explore world population distributions using maps and pie charts. Students also predict and graph world data, such as literacy rates, access to resources, and the distribution of wealth, food, languages, and religions.

National Mathematics Standards (2000): Students in grades 6–8 should "select, create, and use appropriate graphical representations of data" as well as "discuss and understand the correspondence between data sets and their graphical representations" (p. 401). Students should "formulate questions, design studies, and collect data about a characteristic shared by two populations or different characteristics within a population" (p. 401) (Data Analysis and Probability Standard).

National Social Studies Standards (1994): Social studies programs for the middle grades should include experiences that provide for the study of *people, places, and environments* so that the learner can "use appropriate resources, data sources, and geographical tools . . . to generate, manipulate, and interpret information such as atlases, data bases, grid systems, charts, graphs, and maps" (p. 85). They should "distinguish geographic relationships such as population density and spatial distribution patterns" and "propose, compare, and evaluate alternative uses of land and resources in communities, regions, nations, and the world" (p. 85). Social studies programs should include experiences that provide for the study of *global connections and interdependence*, so that the learner can "explore the causes, consequences, and possible solutions to persistent, contemporary, and emerging global issues, such as health, security, resource allocation, economic development, and environmental quality" (p. 102).

Materials: map of the world, bags of one hundred beans for each group of four, rulers, protractors, markers, sticky notes

Description of Activities:

1. Give each group of four students a map of the world and a bag of one hundred precounted beans. Read the first two sentences of *If the World Were a Village* in which the reader learns that the world's population is 6 billion, 200 million. Explain that the one hundred beans in their bags represent the one hundred people in the author's global village. Assist students in understanding that one bean represents 62 million people. Thus, the entire bag of one hundred beans represents 6.2 billion, the world's population. Students place the one hundred beans on various countries and continents as a means to represent their populations.
2. Students share their predictions for global populations. Ask students such questions as, where on your map are most of the beans located? The fewest? What do you think are the populations of these countries, based on the number of beans you have placed on them? Are there continents or countries on which no beans are placed? Which ones?, etc.
3. Share the population data in Smith's book appearing on page 8. By using spreadsheet software or by visiting the Create a Graph Online website, you might prepare in advance a pie chart to provide students with a visual representation of this data. Discuss the accuracy of students' population predictions. Discuss the implications of population densities (especially in such places as India, China, etc.) and the impact on resources, the environment, pollution, and so on.
4. Copy select pages from Smith's book on which a different topic is presented (food, wealth, access to electricity, clean air and water, etc.). Using a sticky note, cover the topic's numerical data, but leave the text introducing and describing the topic uncovered. Distribute one "topic" page to each group of students. Students discuss the topic and generate a pie chart representing their prediction of what the data will look like.
5. Each group presents their pie charts to the class, describing their predictions for their data, after which the teacher reveals the actual data from Smith's book. How accurate were students' predictions? Discuss the implications of each group's topic.
6. Students predict what the global village will look like in the future by exploring the statistics found on page 29 in Smith's book. Students discuss the data, notice the doubling pattern, and then make predictions about the population of the global village (and therefore the world) in the next one hundred years. Students create a line plot showing the population (y-axis) over time (x-axis). Consider exploring the data using a graphing calculator. What are the implications if the population of the world continues to increase at this rate? Facilitate a discussion.

Assessment:

- Did students make reasonable predictions when representing world population distributions?
- Did students notice the unequal distribution of populations worldwide and articulate the implications of this?
- Did students make reasonable predictions about their "topic" data?
- Did students make accurate pie charts and line plots and interpret them correctly?
- Did students articulate the implications of the inequities and disparities in our world?

Activity Extensions:

- Students work in pairs and, using the Internet or a world almanac, locate an interesting world statistic not discussed in *If the World Were a Village*. Students create a pie chart and then share the statistic with the class. For example, students might research and graph blood types, forms of world governance, number of cell phone subscribers or smokers worldwide, and so on.
- Students locate an interesting statistic relative to their town or state, create a pie chart, and then share the statistic with the class. Students might also compare their local statistic to a comparable world statistic and make inferences and draw conclusions. Students might also create an illustration, in the sprit of Smith's book, that captures the statistic.

Cross-Curricular Connections:

Science

- Research and discuss which countries and populations are impacted by malnutrition and famine and the implications of this.
- Research and discuss the history of electricity and its role in advancing society.

Visual Arts

- Explore and discuss the artwork of the twentieth-century U.S. painter and abstract expressionist Richard Diebenkorn, some of whose paintings are aerial landscape views (e.g., *Cityscape I, Berkeley series*, etc.).
- Visit the Six Contemporary Artists Who Use Maps in Their Work website and view, compare, and discuss artists' (Alighiero Boetti, Kathy Prendergast, etc.) renditions of maps.

Related Literature

Garrison, J., & Tubesing, A. (1996). *A million visions of peace: Wisdom from the Friends of Old Turtle.* New York: Scholastic.

Hoose, P. (2002). *It's our world, too!* New York: Farrar Straus Giroux.

Kincade, S. (1992). *Our time is now* (Young people changing the world). Upper Saddle River, NJ: Pearson.

Lewis, B. (1992). *Kids with courage: True stories about young people making a difference.* Minneapolis: Free Spirit.

Pershing Accelerated School Students. (2002). *We dream of a world.* New York: Scholastic.

Roca, N. (2002). *Boys and girls of the world: From one end to the other.* Hauppauge, NY: Barron's Educational Series.

Scillian, D. (2003). *P is for passport.* Chelsea, MI: Sleeping Bear Press.

Smith, D. (2002). *If the world were a village: A book about the world's people.* Tonawanda, NY: Kids Can Press.

Smith, D. (2003). *Mapping the world by heart.* Watertown, MA: Snyder.

Spier, P. (1980). *People.* New York: Doubleday.

Related Instructional Resources

D'Alusio, F. (1998). *Women in the material world.* San Francisco: Sierra Club Books.

Kindersley, A., & Kindersley, B. (1995). *Children just like me.* New York: Dorling Kindersley.

Lewis, B. (1995). *The kid's guide to service projects: Over 500 service ideas for young people who want to make a difference.* Minneapolis: Free Spirit.

Lewis, B. (1998). *The kid's guide to social action: How to solve social problems you choose – and turn creative thinking into positive action.* Minneapolis: Free Spirit.

Menzel, P. (1995). *Material world: A global family portrait.* San Francisco: Sierra Club Books.

Ward, R. (2004). K–8 preservice teachers' journey into the global village: Exploring real-world data using children's literature and technology. *Arizona Reading Journal, 31*(1), 43–47.

Related Websites

Create a Graph Online
http://nces.ed.gov/nceskids/createagraph/

Maps and Map Projections
http://www.btInternet.com/~se16/js/mapproj.htm
http://www.nationalgeographic.com/features/2000/exploration/projections/index.html
http://www.flourish.org/upsidedownmap/

Richard Diebenkorn
http://www.artchive.com/artchive/ftptoc/diebenkorn_ext.html

Six Contemporary Artists Who Use Maps in Their Work
http://www.artjunction.org/articles/mapartists.html

U.S. Census Data and Statistics
http://www.census.gov/
http://www.census.gov/compendia/statab/

Virtual Manipulatives Library—Pie Chart
http://nlvm.usu.edu/en/nav/frames_asid_183_g_2_t_5.html?open=activities

World Data and Statistics
http://www.infoplease.com/
http://www.worldbank.org/data/
http://www.cia.gov/cia/publications/factbook/
http://www.wri.org/
http://www.unicef.org/
http://www.un.org/
http://plasma.nationalgeographic.com/places/index.html
http://www.pbs.org/wgbh/nova/worldbalance/eart-nf.html
http://mapping.com/gv/
http://www.prb.org/
http://www.readtofeed.org/

The Sky's the Limit: Stories of Discovery by Women and Girls (2002)

by Catherine Thimmesh

Houghton Mifflin, ISBN #0618494898

Mathematicians Are People, Too (1995)

by Luetta Reimer and Wilbert Reimer

Dale Seymour, ISBN #0866518231

Five Brilliant Scientists (Great Black Heroes Series) (2000)

by Lynda Jones

Scholastic, ISBN #0590480316

Overview of Books: In *The Sky's the Limit: Stories of Discovery by Women and Girls,* discover the amazing discoveries and stories of women who have made our world a better, safer, and more technologically advanced place. Then, in *Mathematicians Are People, Too,* enjoy fifteen stories about the lives of such great mathematicians as Fibonacci, Rene Descartes, Sonya Kovalesky, Ada Lovelace, George Polya, and others. Celebrate the work and accomplishments of five individuals in *Five Brilliant Scientists* (Great Black Heroes Series), which contains brief biographies on Susan McKinney Steward (the first African American female doctor in New York), George Washington Carver (agricultural scientist), Ernest Everett Just (marine biologist), Percy Lavon Julian (chemist), and Shirley Ann Jackson (nuclear scientist).

Mathematical Concepts and Skills: real-life applications of mathematics, careers in mathematics

Social Studies Concepts and Skills: identification of scientific and technological discoveries and advances, careers, biography as a form of storytelling

Overview of Activities: Students read about and discuss the various mathematical, scientific, and technological discoveries of and inventions made by men and women. Students then create a trifold poster featuring illustrations and highlights of a particular individual's life and achievements. Students rank, debate, and discuss the importance of each individual's discoveries and work in terms of its impact on advancing and bettering society.

National Mathematics Standards (2000): Students in grades 6–8 should "recognize and apply mathematics in contexts outside of mathematics" (p. 402) (Connections Standard). Students should also "communicate their mathematical thinking coherently and clearly to peers, teachers, and others" and "use the language of mathematics to express mathematical ideas precisely" (p. 402) (Communication Standard).

National Social Studies Standards (1994): Social studies programs for the middle grades should include experiences that provide for the study of *relationships among science, technology, and society,* so that the learner can "examine and describe the influence of culture on scientific and technological choices and advancement, such as transportation, medicine, and warfare" and "show through specific examples how science and technology have changed people's perceptions of the social and natural world, such as in their relationship to the land, animal life, family life, and economic needs, wants, and security" (p. 99).

Materials: Internet or other informational resources, trifold poster

Description of Activities:

1. Place students into pairs and assign them (or let students select) the name of one of the scientists or inventors appearing in *The Sky's the Limit: Stories of Discovery by Women and Girls; Mathematicians Are People, Too;* or in *Five Brilliant Scientists.*
2. Using the information appearing in the featured books as a springboard, as well as Internet or library resources, students create a trifold poster featuring biographical information, graphics, and highlights of their individual's life and achievements. Students should include a description of the mathematical knowledge and other content knowledge and skills the individuals used or needed as they worked on their research, discoveries, and inventions.
3. Students share their posters with the class. Students rank, debate, and discuss the importance of each individual's discoveries and work in terms of its impact on advancing and bettering society.

Assessment:

- Did students capture the main highlights and accomplishments of the individual they researched?
- Did students clearly articulate the mathematical and other content knowledge and skills the individuals used or needed as they worked on their research, discoveries, and inventions?

Activity Extensions:

- Read the article posted on the *Newsweek*—The Power of Big Ideas website listed below and let students debate which invention was the most important in the last two thousand years.
- Facilitate a discussion regarding students' choices of future careers, with a focus on the math skills they will need to know to perform well in this job. Encourage students to discuss how this job contributes to society.
- Visit the U.S. Patent and Trademark Office website and research the patent process, how many patents are in existence, who has received the largest number of patents, the first women to receive a patent, and other facts. Also, research the difference between a patent, trademark, and copyright.

Cross-Curricular Connections:

Science

- Create a timeline of advances in medicine, science, or technology.
- Research and discuss the inventions of Leonardo da Vinci.

Visual Arts

- Create a collage containing images of individuals engaged in careers that necessitate knowledge of mathematics or the sciences.

Related Literature

Anderson, M. (1995). *Amazing Leonardo da Vinci inventions you can build yourself.* Chicago: Nomad Books.

Brewer, D. (2004). *Inventions* (100 things you should know about series). New York: Barnes & Noble Books.

Haber, L. (1992). *Black pioneers of science and invention.* San Diego: Harcourt Children's Books.

Jones, L. (2000). *Five brilliant scientists* (Great black heroes series). New York: Scholastic.

LaFontaine, B. (1998). *Great inventors and inventions* (100 things you should know about series). London: Dover.

Reimer, L., & Reimer, W. (1995). *Mathematicians are people, too.* Palo Alto, CA: Dale Seymour.

Thimmesh, C. (2000). *Girls think of everything: Stories of ingenious inventions by women.* Boston: Houghton Mifflin.

Thimmesh, C. (2002). *The sky's the limit: Stories of discovery by women and girls.* Boston: Houghton Mifflin.

Related Instructional Resources

Inventor and Inventions Series (Benchmark Books)

Chipman, D. (1998). *Cool women.* Chicago: Girl Press.

Egan, L. (1999). *Inventors and inventions.* New York: Scholastic.

Hahn, E. (1996). *16 extraordinary American women.* Portland, ME: J. Weston Walch.

Flack, J. (1989). *Inventing, inventions, and inventors: A teaching resource book.* Englewood, CO: Teacher Ideas Press.

Kassinger, R. (2002). *Build a better mousetrap: Make classic inventions, discover your problem-solving genius, and take the inventor's challenge.* Hoboken, NJ: Wiley & Sons.

Saunders, H. (1988). *When are we ever gonna have to use this?* Palo Alto, CA: Dale Seymour.

VanCleave, J. (2004). *Scientists through the ages.* Hoboken, NJ: Wiley & Sons.

Welden, A. (1998). *Girls who rocked the world: Heroines form Sacagawea to Sheryl Swoopes.* Hillsboro, OR: Beyond Words.

Related Websites

Biographies of Famous Mathematicians

http://www-groups.dcs.st-and.ac.uk/~history/BiogIndex.html

Biographies of Scientists

http://www.blupete.com/Literature/Biographies/Science/Scients.htm
http://www.infoplease.com/spot/scibio8.html
http://www.maths.tcd.ie/pub/HistMath/Links/SciBiog.html

Biographies of Women in Science

http://www.awm-math.org/biographies.html#sciencebios

Great Inventors

http://library.thinkquest.org/5847/homepage.htm
http://library.christchurch.org.nz/Kids/Inventions/WhoInvented/
http://www.enchantedlearning.com/inventors/medicine.shtml

Newsweek—**The Power of Big Ideas**

http://antiquespectacles.com/history/images/Newsweek-150.pdf

U.S. Patent and Trademark Office

http://www.uspto.gov/

Women Inventors

http://inventors.about.com/library/blwomeninventors.htm
http://inventors.about.com/od/womeninventors/index_a.htm
http://www.ideafinder.com/features/classact/women.htm

3

Literature-Based Mathematics and Visual Arts Activities

The Mathematics-Visual Arts Connection

The arts (visual art, music, movement, and drama) present an ideal and unique forum through which students can express their ideas, thoughts, and emotions. In particular, the visual arts, which range from "drawing, painting, sculpture, and design, to architecture, film, video, and folk arts" (MENC, 1994, p. 33), embody mathematics, as students can explore such interrelated concepts as patterns, line, shape, and form. While learning the mathematical concepts embedded in the visual arts, students can collaboratively engage in communicating, reasoning, and investigating, activities that both the NCTM (1989, 2000) and MENC (1994) strongly advocate. A growing body of evidence documents that learning in the arts involves principles

shared with other academic disciplines (Bransford et al., 2004; Deasy, 2002; Fiske, 1999; Scripp, 2002). In his book *Arts with the Brain in Mind,* Jensen (2001) contends that the "visual arts seem to be strongest when used as a tool for academic learning" (p. 58). Further, Eisner (1998, 2004) cites many studies that report strong links between visual learning and improvement in reading and creativity. Thus, it is not surprising that integrating the arts with other content areas is mounting (Bickley-Green, 1995; Efland, 2002; McDonald & Fisher, 2006; Phillips & Bickley-Green, 1998; Walling, 2005; Ward, 2006a). Further, many pieces of literature, such as biographies and poetry, can bring the learning of mathematics, as well as other content areas, and the visual arts to life (McDonald & Fisher, 2006; Ward, 2004a, 2005).

"The visual arts embody interrelated mathematics concepts such as patterns, line, shape, and form."

This chapter articulates a variety of literature-based activities that integrate concepts and skills used and learned in the study of mathematics with those in the visual arts. While engaged in these activities, students will discover and gain practice with such mathematics concepts and skills as fractions, decimals, percents, ratio, proportion (Number and Operations Standard); algebra, algebraic equations, balancing equations, solving for one unknown, pattern recognition (Algebra Standard); tessellations, transformations, lines, angles, two- and three-dimensional shapes, nets, Platonic Solids, faces, edges, vertices, spatial reasoning (Geometry Standard); size, scale, angle measurement, volume (Measurement Standard); probability, random events, likelihood, fair chance, theoretical probability, experimental probability, Law of Large Numbers (Data Analysis and Probability Standard); and the four-color map theorem and graph theory (Problem Solving Standard).

Art concepts and skills featured in this chapter include line, shape, color, value, texture, form, space, art history, and using various art medium. While exploring mathematics, students will simultaneously learn about and mimic the work of such artists including Hans (Jean) Arp, Alexander Calder, Paul Cezanne, M. C. Escher, Paul Klee, Roy Lichtenstein, Joan Miro, Piet Mondrian, Georgia O'Keeffe, Pablo Picasso, and others.

The integrated literature-based activities also provide students with many opportunities to predict, estimate, problem-solve, and reason (Problem

Solving and Reasoning and Proof Standards) as well as communicate and use various representations to organize, record, model, and interpret mathematical ideas (Communication and Representations Standards). Further, students will also discover and explore real-life applications of mathematics and the visual arts and careers in mathematics and the visual arts (Connections Standard).

Remember to check the appendix for ideas and samples of assessment tools and rubrics.

Matrix of Mathematics and Visual Arts Activities

BOOK TITLE	MATHEMATICAL CONCEPTS AND SKILLS	VISUAL ARTS CONCEPTS AND SKILLS	SCIENCE CONCEPTS AND SKILLS	SOCIAL STUDIES CONCEPTS AND SKILLS
Roy Lichtenstein: The Artist at Work; Roy Lichtenstein (Getting to Know the World's Greatest Artists Series)	fractions, decimals, percents	line, shape, color, creating art (pop art using Benday-like dots), art history (life and artwork of Roy Lichtenstein)	investigate the process of creating Benday dots	explore and discuss a biography of other pop artists
Georgia's Bones; Georgia O'Keeffe (Getting to Know the World's Greatest Artists Series); *Georgia O'Keeffe: One Hundred Flowers*	size, scale, ratio, proportion, measurement	line, shape, color, value, space, creating art (scaled O'Keeffe-like flowers), art history (life and artwork of Georgia O'Keeffe)	investigate the reproductive process of flowering plants	explore and discuss a biography of other abstract artists; research and discuss the geography of New Mexico
Alexander Calder (The Life and Work of . . . Series); *Calder* (Great Modern Masters Series)	algebra, algebraic equations, solving for one unknown, balancing equations	line, shape, color, form, space, creating art (balancing mobiles), art history (life and artwork of Alexander Calder)	research and discuss the many inventions and war machines of Archimedes	research and discuss careers in art or in engineering
Picasso (Famous Artists Series); *Pablo Picasso: Breaking All the Rules* (Smart about Art Series)	Platonic Solids, two- and three-dimensional shapes, spatial reasoning, nets, faces, edges, vertices	line, shape, form, space, creating art (cubist artwork), art history (life and artwork of Pablo Picasso)	explore and discuss the Platonic Solids' associations with the elements in the universe	explore and discuss a biography of other cubist artists
Tessellations: The History and Making of Symmetrical Designs; M. C. Escher: The Graphic Work	tessellations, patterns, transformations (translations, rotations, reflections, glide reflections)	line, shape, creating art (tessellations), art history (life and artwork of M. C. Escher)	find examples of tessellations in nature; investigate how and why tessellations are used in architecture	look for tessellations in woven patterns of Native American rugs and quilts; explore tessellations in Islamic architecture

BOOK TITLE	MATHEMATICAL CONCEPTS AND SKILLS	VISUAL ARTS CONCEPTS AND SKILLS	SCIENCE CONCEPTS AND SKILLS	SOCIAL STUDIES CONCEPTS AND SKILLS
Mondrian (Great Modern Masters Series)	parallel lines, perpendicular lines, intersecting lines, transversal, angles (vertical, supplementary, complementary, alternate interior, alternate exterior, corresponding, consecutive interior), angle measurement, angle relationships	line, shape, color, creating art (abstract art), art history (life and artwork of Piet Mondrian)	investigate and discuss chromatics, identify angles occurring in nature	explore and discuss a biography of other abstract artists
Cezanne (Famous Artists Series); *Paul Cezanne: A Painter's Journey*	three-dimensional shapes, cones, cylinders, pyramids, prisms, volume	line, shape, color, value, form, space, creating art (still life), art history (life and artwork of Paul Cezanne)	explore and discuss the engineering knowledge and skills required to build pyramids	explore and discuss a biography of artists influenced by Cezanne
Miro: Earth and Sky (Art for Children Series); *Joan Miro* (Famous Artists Series)	probability, random events, likelihood, fair chance, theoretical probability, experimental probability, Law of Large Numbers, prediction	line, shape, color, form, space, creating art (surrealist art), art history (life and artwork of Joan Miro)	explore and discuss stars and constellations	research and discuss the geography and demographics of the Catalonia region of Spain
Arp; *Looking at Dada*	probability, random events, Law of Large Numbers, prediction, pi	line, shape, color, value, texture, space, creating art (chance collage), art history (life and artwork of Jean Arp and Dada)	research and discuss the role of probability in genetics	research and discuss the causes, major events, duration, and aftermath of World War I
Paul Klee (The Life and Work of . . . Series); *Paul Klee* (Getting to Know the World's Greatest Artists Series)	four-color map theorem, problem solving, estimation, graph theory	line, shape, color, value, creating art (watercolored name strip), art history (life and artwork of Paul Klee)	locate areas on a map that have experienced constructive and destructive forces	identify various geographic features on a map

Roy Lichtenstein: The Artist at Work (1994)

by Lou Ann Walker

Lodestar Books, ISBN #0525674357

Roy Lichtenstein (Getting to Know the World's Greatest Artists Series) (2002)

by Mike Venezia

Children's Press, ISBN #0516259636

Overview of Books: In *Roy Lichtenstein: The Artist at Work,* visit the art studio of the U.S. twentieth-century pop artist Roy Lichtenstein, famous for his use of Benday dots, and witness how he develops his ideas into artwork. Then, enjoy more facts and photos about the life and artwork of Roy Lichtenstein in the biography, *Roy Lichtenstein* (Getting to Know the World's Greatest Artists Series).

Mathematical Concepts and Skills: fractions, decimals, percents

Visual Arts Concepts and Skills: line, shape, color, creating art (pop art using Benday-like dots), art history (life and artwork of Roy Lichtenstein)

Overview of Activities: Students explore and learn about the life and artwork of Roy Lichtenstein and visually estimate and then compute fractions and percents using images of his artwork. Students then create a piece of pop art in the spirit of Lichtenstein.

National Mathematics Standards (2000): Students in grades 6–8 should "work flexibly with fractions, decimals, and percents to solve problems" (p. 393). Students should also "select appropriate methods and tools for computing with fractions and decimals from among mental computation, estimation, calculators or computers, and paper-and-pencil, depending on the situation, and apply the selected methods" (p. 393) (Number and Operations Standard).

National Visual Arts Standards (1994): Students in grades 5–8 should use knowledge of structures and functions (Content Standard 2) and "generalize about the effects of visual structures and functions and reflect upon these effects in their own work" (p. 50) (Achievement Standard a); choose and evaluate a range of subject matter, symbols, and ideas (Content Standard 3) and "integrate visual, spatial, and temporal concepts with content to communicate intended meaning in their artworks" (p. 50) (Achievement Standard a); reflect upon and assess the characteristics and merits of their work and the work of others (Content Standard 5) and "describe and compare a variety of individual responses to their own artworks and to artworks from various eras and cultures" (p. 51) (Achievement Standard c); make connections between visual arts and other disciplines (Content Standard 6) and "describe ways in which the principles and subject matter of other disciplines taught in the school are interrelated to the visual arts" (p. 51) (Achievement Standard b).

Materials: images of Lichtenstein's works (e.g., *Whaam!, Girl with Hair Ribbon, Drowning Girl, Red Barn, Baked Potato, Sponge II, Explosion*), newspaper comic strip, magnifying glass, overhead copy of centimeter paper, 9" × 12" white paper, markers, rulers

Description of Activities:

1. Share with students select passages about the life and artwork of Roy Lichtenstein using the books *Roy Lichtenstein: The Artist at Work* and *Roy Lichtenstein*. Roy Lichtenstein, a twentieth-century U.S. artist, is best known for initiating a period in art known as pop art, which is art that depicts everyday or popular images. Many of Lichtenstein's works look like comic book pages because of his use of Benday dots. Benday dots is a printing process named after its inventor, Benjamin Day, in which colored dots (cyan, magenta, yellow, and black) are closely spaced, widely spaced, or overlap to create a shading effect as well as such secondary colors including green, purple, orange, and flesh tones.
2. Give students a newspaper comic strip and a magnifying glass to allow them to view Benday dots. Benday dots were considered the hallmark of artist Roy Lichtenstein, who enlarged and exaggerated them in many of his paintings and sculptures. Show images of Lichtenstein's work that feature his Benday dots (e.g., *Whaam!, Girl with Hair Ribbon, Explosion, Drowning Girl,* etc.).

3. Place students in small groups. Distribute color images of Lichtenstein's work (e.g., *Red Barn, Baked Potato, Sponge II, Girl with Hair Ribbon, Explosion*). Ask students to first visually estimate what fractional part of the painting is comprised of particular colors, dots, and line patterns (e.g., What fractional part of *Red Barn* is red? Is yellow? Is dotted?, etc.). Students convert the fractions to percents. Students can verify their work by checking that the fractions add to 1 and that the percents sum to 100 percent.
4. Distribute an overheard copy of centimeter graph paper to students. Students place the clear grid on top of the Lichtenstein color image and first count the total number of squares covering the entire color image. Students then count and compute what fractional part of the painting is comprised of particular colors, dots, and line patterns.
5. Students compare their estimations (from step 3) to their computed values (from step 4). How accurate were their estimations?
6. Students identify a familiar, everyday object and create a piece of artwork in the spirit of Lichtenstein, using dots and bold colors. Read excerpts from pages 7–17 in *Roy Lichtenstein: The Artist at Work,* which describe how Lichtenstein creates his artwork using a ruler and black tape. Consider using a dot stencil to help students create their Benday dots.
7. Students share their artwork with the class. Students discuss whose artwork is most visually appealing and why. Whose artwork embodies the spirit of Lichtenstein the most?

Assessment:

- Did students accurately estimate and then compute the fractional amount of each color, lined, or dot pattern appearing in a Lichtenstein image?
- Did students correctly convert all fractions to percents?
- Did students create their own piece of artwork in the spirit of Lichtenstein?
- Did students articulate the features they liked and disliked in the class's artwork?

Activity Extensions:

- Students create a pie chart representing the percentage of colors, lined, or dot patterns appearing in a Lichtenstein color image.
- Students estimate the total number of dots appearing in a Lichtenstein artwork.

- Show examples of the artwork of other pop artists, in particular, the works of Andy Warhol. Facilitate a discussion about what features students find appealing (or unappealing) about pop artists' works.
- Explore the art period known as pointillism and the works of George Seurat. Estimate the number of dots of paint used in one of his paintings (e.g., *Sunday Afternoon on the Island of La Grand Jatte*).

Cross-Curricular Connections:

Science

- Investigate the process of creating Benday dots.

Social Studies

- Explore and then share findings of a biographical piece of a pop artist (e.g., Andy Warhol, Robert Rauschenberg, John Chamberlain, etc.).

Related Literature

Alloway, L. (1983). *Roy Lichtenstein.* New York: Abbeville Press.

Appel, J., & Guglielmo, A. (2006). *Pop Warhol's top.* New York: Sterling.

Clements, A. (2006). *A million dots.* New York: Simon & Schuster Books for Young Readers.

Hickey, D., & Brown, E. (2005). *Roy Lichtenstein: Prints from the Jordan D. Schnitzer Collection.* Seattle: Marquand Books.

Venezia, M. (2002). *Georges Seurat* (Getting to know the world's greatest artists series). New York: Children's Press.

Venezia, M. (2002). *Roy Lichtenstein* (Getting to know the world's greatest artists series). New York: Children's Press.

Walker, L. (1994). *Roy Lichtenstein: The artist at work.* New York: Lodestar Books.

Related Instructional Resources

Aigner-Clark, J. (2004). *Baby Einstein: The ABCs of art.* New York: Hyperion Books for Children.

Dickins, R. (2005). *The children's book of art: An introduction to famous paintings.* London: Usborne.

Evans, J., & Skelton, T. (2001). *How to teach art to children.* Monterey, CA: Evan-Moor.

Kohl, M., & Solga, K. (1996). *Discovering great artists: Hands-on art for children in the styles of the great masters.* Bellingham, WA: Bright Ring.

Krull, K. (1995). *Lives of the artists: Masterpieces, messes.* San Diego: Harcourt Brace.

Renshaw, A., & Ruggi, G. (2005). *The art book for children.* New York: Phaidon Press.

Scieszka, J., & Smith, L. (2005). *Seen art?* New York: Viking Press.

Related Websites

Benday Dots
http://www.henryart.org/ex/lichtenstein.htm

History of Art Timeline
http://www.printsforpleasure.com/historyofart.htm

Major Modern and Contemporary Visual Artists
http://www.the-artists.org/index.cfm

Pop Art
http://www.sanford-artedventures.com/study/g_pop_art.html

Roy Lichtenstein
http://www.lichtensteinfoundation.org/
http://www.artcyclopedia.com/artists/lichtenstein_roy.html

Study Art
http://www.sanford-artedventures.com/study/study.html

Teach Art
http://www.sanford-artedventures.com/teach/teach.html

Georgia's Bones (2005)

by Jen Bryant

Eerdmans Books for Young Readers, ISBN #0802852173

Georgia O'Keeffe (Getting to Know the World's Greatest Artists Series) (1993)

by Mike Venezia

Children's Press, ISBN #0516422979

Georgia O'Keeffe: One Hundred Flowers (1989)

by Georgia O'Keeffe

Knopf, ISBN #0760711127

Overview of Books: Enjoy a story about the life of twentieth-century U.S. artist Georgia O'Keeffe and her fascination with shapes from her young days on a Wisconsin farm to her adult years in New York and New Mexico in *Georgia's Bones*. In *Georgia O'Keeffe* (Getting to Know the World's Greatest Artists Series), learn about the life and colorful artwork of O'Keeffe, most famous for her boldly colorful artwork featuring shells, animal bones, landscapes, and close-ups of flowers. Then, in the oversized book, *Georgia O'Keeffe: One Hundred Flowers,* view one hundred flowers painted by O'Keeffe.

Mathematical Concepts and Skills: size, scale, ratio, proportion, measurement

Visual Arts Concepts and Skills: line, shape, color, value, space, creating art (scaled O'Keeffe-like flowers), art history (life and artwork of Georgia O'Keeffe)

Overview of Activities: Students explore and learn about the life and artwork of Georgia O'Keeffe, who used bold colors to paint close-ups of flowers. Students determine whether the flowers in particular O'Keeffe paintings are drawn to scale by measuring live, similar flowers. Students sketch and color a scaled drawing of a flower in the spirit of Georgia O'Keeffe.

National Mathematics Standards (2000):

Students in grades 6–8 should "develop, analyze, and explain methods for solving problems involving proportions, such as scaling and finding equivalent ratios" (p. 393) (Number and Operations Standard). Students in grades 6–8 should "solve problems involving scale factors, using ratio and proportion." Students should also "select and apply techniques and tools to accurately find length, area, volume, and angle measures to appropriate levels of precision" (p. 399) (Measurement Standard).

National Visual Arts Standards (1994):

Students in grades 5–8 should use knowledge of structures and functions (Content Standard 2) and "employ organizational structures and analyze what makes them effective or not effective in the communication of ideas" (p. 50) (Achievement Standard b); reflect upon and assess the characteristics and merits of their work and the work of others (Content Standard 5) and "describe and compare a variety of individual responses to their own artworks and to artworks from various eras and cultures" (p. 51) (Achievement Standard c); make connections between visual arts and other disciplines (Content Standard 6) and "describe ways in which the principles and subject matter of other disciplines taught in the school are interrelated to the visual arts" (p. 51) (Achievement Standard b).

Materials:

images of O'Keefe's works (e.g., *Purple Petunia, Morning Glory with Black, Pink Tulip, Red Poppy, White Rose,* etc.), live flowers depicted in O'Keeffe's paintings, rulers, 9" × 12" posterboard, coloring medium (colored pencils, crayons, oil paints, or watercolors)

Description of Activities:

1. Share with students select passages about the life and artwork of Georgia O'Keeffe from *Georgia's Bones* and *Georgia O'Keeffe.* Share the quote spoken by O'Keeffe in the opening pages of *Georgia O'Keeffe: One Hundred Flowers* in which she comments on the small size of a flower and, if she were to paint it true to its size, no one would see it. Let students view several examples of O'Keeffe's artwork from her book (e.g., *Red Canna* [plate 15], *Petunia* [plate 21], *Morning Glory with Black* [p. 39], *Calla Lillies with Red Anemone* [plate 56], etc.). Encourage students to see how her flowers fill the canvas, with petals going off the canvas's edges, making viewers feel as though they were deep inside of the flower.
2. Share other pieces of O'Keeffe's artwork appearing in *Georgia O'Keeffe* and *Georgia O'Keeffe: One Hundred Flowers,* including those of animal bones (*Cow's Skull with Calico Rose*), shells (*White Shell with Red*), and mountains and landscapes (*Grey Hills*). Facilitate a discussion about what features students find appealing (or unappealing) about her artwork.

3. Set up centers in the classroom. At each center, place a color image of an O'Keeffe flowered work (e.g., *Purple Petunia, Morning Glory with Black, Pink Tulip, Red Poppy, White Rose,* etc.), rulers, and the actual live flower depicted in the painting. Students sit four per center and work in pairs. Students observe and compare the live flower at their center to the artist's rendering. Students make a scaled sketch of the flower and measure and record its parts. Students measure O'Keeffe's painted flower and its parts. Students compute the scale factor used to transform the small live flower into O'Keeffe's painted flower. Did both pairs compute the same scale factor? Encourage students to explain their strategies for finding the scale factor. Students also discuss whether O'Keeffe scaled her flowers accurately.
4. Using 9" × 12" posterboard, coloring medium, and their pencil sketch, students use a larger scale factor (as compared to O'Keeffe) and sketch and color the flower.
5. Students share their artwork with the class, comparing it to O'Keeffe's rendering. How accurate were students in scaling their flowers? What strategies did students use to scale their drawings?

Assessment:

- Did students accurately sketch, measure, and draw to scale their live flower?
- Did students accurately estimate the scale factor used in the O'Keeffe image?
- Did students clearly articulate the strategies they used in scaling their work?

Activity Extensions:

- Students estimate and then create a pie chart representing the percentage of colors appearing in an O'Keeffe color image.
- View O'Keeffe's *White Shell with Red*. Discuss how the shell's spiral is a "golden spiral" (i.e., a spiral formed by making successively smaller golden rectangles). Visit the Golden Spiral website for more information.
- Students explore the life and work of the Swedish botanist, Carl Linnaeus, who sketched the minute details of flower parts and plants and then hired artists to illustrate his findings and work.

Cross-Curricular Connections:

Science

- Investigate the reproductive process of flowering plants.

Social Studies

- Explore and then share findings of a biographical piece of another abstract artist (e.g., Georges Braque, Juan Gris, etc.).
- Research and discuss the geography of New Mexico, in particular, Ghost Ranch, where O'Keeffe retired later in life and whose painted desert and enormous cliffs and hills inspired some of her most famous landscapes.

Related Literature

Brooks, P. (1995). *Georgia O'Keeffe: An adventurous spirit.* Chicago: Franklin Watts.

Bryant, J. (2005). *Georgia's bones.* Grand Rapids, MI: Eerdmans Books for Young Readers.

Greenberg, J., & Jordan, S. (1994). *The American eye: Eleven artists of the twentieth century.* New York: Delacorte Press.

Lowery, L. (1996). *Georgia O'Keeffe.* Minneapolis: Carolrhoda Books.

Venezia, M. (1993). *Georgia O'Keeffe* (Getting to know the world's greatest artists series). Chicago: Children's Press.

Related Instructional Resources

Aigner-Clark, J. (2004). *Baby Einstein: The ABCs of art.* New York: Hyperion Books for Children.

Dickins, R. (2005). *The children's book of art: An introduction to famous paintings.* London: Usborne.

Evans, J., & Skelton, T. (2001). *How to teach art to children.* Monterey, CA: Evan-Moor.

Kohl, M., & Solga, K. (1996). *Discovering great artists: Hands-on art for children in the styles of the great masters.* Bellingham, WA: Bright Ring.

Krull, K. (1995). *Lives of the artists: Masterpieces, messes.* San Diego: Harcourt Brace.

O'Keeffe, G. (1989). *Georgia O'Keeffe: One hundred flowers.* New York: Knopf.

Renshaw, A., & Ruggi, G. (2005). *The art book for children.* New York: Phaidon Press.

Scieszka, J., & Smith, L. (2005). *Seen art?* New York: Viking Press.

Related Websites

Georgia O'Keeffe

http://www.princetonol.com/groups/iad/lessons/middle/okeeffe.htm
http://gardenofpraise.com/art16.htm
http://www.artcyclopedia.com/artists/okeeffe_georgia.html
http://www.okeeffemuseum.org/indexflash.php

Golden Spiral

http://britton.disted.camosun.bc.ca/goldengeom/goldengeom2.html

History of Art Timeline

http://www.printsforpleasure.com/historyofart.htm

Major Modern and Contemporary Visual Artists

http://www.the-artists.org/index.cfm

Study Art

http://www.sanford-artedventures.com/study/study.html

Teach Art

http://www.sanford-artedventures.com/teach/teach.html

Activities Featuring Algebra

Alexander Calder (The Life and Works of . . . Series) (2003)

by Adam Schaefer

Heinemann Library, ISBN #1403404933

Calder (Great Modern Masters Series) (1997)

by Gerard-Georges Lemaire

Abrams, ISBN #0810946688

Overview of Books: In both of these books, learn about the works of the U.S. artist and sculptor, Alexander "Sandy" Calder, best known for his creation of giant mobiles and stabiles.

Mathematical Concepts and Skills: algebra, algebraic equations, solving for one unknown, balancing equations

Visual Arts Concepts and Skills: line, shape, color, form, space, creating art (balancing mobiles), art history (life and artwork of Alexander Calder)

Overview of Activities: Students explore and learn about the life and artwork of Alexander "Sandy" Calder, famous for his large-scale mobiles. Students engage in an experiment with pennies and Popsicle sticks to discover the Law of Levers, used by Calder to balance his mobiles. Students then balance and create their own mobile in the spirit of Calder.

National Mathematics Standards (2000): Students in grades 6–8 should "represent, analyze, and generalize a variety of patterns with tables, graphs, words, and, when possible, symbolic rules" (p. 395). Students should "use symbolic algebra to represent situations and to solve problems, especially those that involve linear relationships" (p. 395) (Algebra Standard).

National Visual Arts Standards (1994): Students in grades 5–8 should use knowledge of structures and functions (Content Standard 2) and "generalize about the effects of visual structures and functions and reflect upon these effects in their own work" (p. 50) (Achievement Standard a); choose and evaluate a range of subject matter, symbols, and ideas (Content Standard 3) and "integrate visual, spatial, and temporal concepts with content to communicate intended meaning in their artworks" (p. 50) (Achievement Standard a); reflect upon and assess the characteristics and merits of their work and the work of others (Content Standard 5) and "describe and compare a variety of individual responses to their own artworks and to artworks from various eras and cultures" (p. 51) (Achievement Standard c); make connections between visual arts and other disciplines (Content Standard 6) and "describe ways in which the principles and subject matter of other disciplines taught in the school are interrelated to the visual arts" (p. 51) (Achievement Standard b).

Materials: images of Calder's *Cirque Calder*, images of Calder's mobiles (*Lobster Trap and Fish Tail, Ghost, International Mobile*, etc.), images of Calder's stabiles, (e.g., *Five Swords, Flamingo, Carmen*, etc.), images of Calder's standing mobiles (e.g., *Indian Feathers, Bird on the Tree Branch, Two Moons*, etc.), Popsicle sticks, bags of pennies, rulers, pencils, calculators, scales, string, dowel rods, foam stickers, everyday items (to be used in the creation of the mobile), *Alexander Calder* worksheet

Description of Activities:

1. Share with students select passages about the life and artwork of Alexander Calder from *Alexander Calder* (The Life and Work of . . . Series) and *Calder* (Great Modern Masters Series). Alexander Calder was a twentieth-century U.S. artist who not only painted and sculpted, but also created lithographs and designed tapestries and carpets. Show students an image of one of Calder's most famous works, his five-year project entitled *Cirque Calder*, which was a miniature circus of characters and animals made of wire, scraps of cloth, wood, cork, and other materials.
2. Explain to students how Calder began dabbling in the art form of mobile construction around 1930, utilizing his extensive mathematics and engineering background. Calder's mobiles are moving, three-dimensional, abstract sculptures consisting of shapes suspended from lightweight rods or wire. Show students some of Calder's mobiles (e.g., *Lobster Trap and Fish Tail, Ghost, International Mobile*, etc.). Facilitate a discussion in which student articulate what they notice, like, and dislike about Calder's art. Do his mobiles appear balanced? Do students like the visual asymmetry in his mobiles?
3. At first glance, many of Calder's mobiles may appear unbalanced because one side looks heavier than the other. However, Calder discovered that balance could be achieved if the weight of the objects suspended multiplied by their distance from the pivot (or balancing) point on the left-hand side of the mobile equaled the product of the weight of the objects suspended and their distance from the pivot point on the right-hand side of the mobile (i.e., $W_L \times D_L = W_R \times D_R$). This relationship is also known as the Law of Levers, originally formulated by the great Greek mathematician, Archimedes, around 260 B.C.E.

4. Challenge students to discover the Law of Levers and the variables involved in this formula. Give students a bag of thirty pennies, a pencil, and a Popsicle stick. Ask students to balance the Popsicle stick on the pencil by placing them perpendicular to one another. Students will see that the Popsicle stick balances when the pencil is placed at the center (fulcrum) of the Popsicle stick. Students place one penny on the left-hand side of the stick. Students will see that the lever (Popsicle stick) is no longer balanced. Encourage students to articulate that weight is a factor in creating balance. Keeping one penny on the left-hand side of the pencil (lever), students slide the pencil away from the center of the Popsicle stick until balance is achieved. Encourage students to articulate that the distance to the fulcrum (balancing point) must be a factor in creating balance.

5. To discover how weight and distance on both sides of the lever must interact, students work in pairs and perform an experiment in which they weigh a stack of pennies and then place them on the left-hand side of the Popsicle stick. Students weigh another stack of pennies (using a different number of pennies) and place them on the right-hand side of the Popsicle stick. Students slide the stacks of pennies across the Popsicle stick (or move the pencil, i.e., the fulcrum, underneath the Popsicle stick) until balance is achieved. Students then measure the distance from the center of both stacks of pennies to the fulcrum (balancing point) and record these values on the *Alexander Calder* worksheet. Students repeat this step five or six times, recording the weights of the stacks of pennies and distances on both sides of the Popsicle stick each time.

6. Students examine their table of data and look for a pattern. Assist students in determining how weight and distance on the left-hand side of the lever must be manipulated such that they equal (or balance) the weight and the distance on the right-hand side of the lever. Students will find that if you multiply the weight and distance on the left-hand side of the lever, this will yield the same product as the weight multiplied by the distance on the right-hand side of the lever. Introduce to students this formula, known as the Law of Levers, first derived by Archimedes, and used by Calder to balance his mobiles: $W_L \times D_L = W_R \times D_R$. Explain what each variable and subscript represents.

7. In groups of four, students work collaboratively to create a two- or three-tier mobile in the spirit of Calder using dowel rods of varying lengths, string, and everyday items such as bolts, plastic utensils, paper clips, foam stickers, etc., serving as the objects that are suspended from the string. In building their mobiles, students weigh the objects on both sides of the balancing point and verify that the Law of Levers does indeed hold. Also, as did Calder, students should carefully select and suspend at specific distances the objects in the mobile, indicative of the meaning behind the mobile. Students title their mobile.
8. Students examine their classmates' mobiles, noticing how they achieved balance and articulating the appeal of each mobile.

Assessment:

- Did students accurately record weights and distances in their table of values?
- Did students discover the Law of Levers?
- Did students work collaboratively to build and balance a Calder-like mobile?
- Did students articulate what they liked and disliked about each group's mobiles?

Activity Extensions:

- Provide students with three of the four variables in the Law of Levers. Students learn the steps of solving equations with one unknown.
- Using wire or pipe cleaners, students create a Calder-like masterpiece in the spirit of his *Cirque Calder*.
- Show students images of Calder's "stabiles," big sculptures that, unlike mobiles, did not move (e.g., *Five Swords, Flamingo, Carmen*, etc.). Calder designed his stabiles such that they looked different from different directions as people walked around them. Calder also combined mobiles and stabiles, creating what he called standing mobiles (e.g., *Indian Feathers, Bird on the Tree Branch, Two Moons*, etc.).

Cross-Curricular Connections:

Science

- Research and discuss the many inventions and war machines of Archimedes.

Social Studies

- Research and discuss careers in art or in engineering.

Related Literature

Lemaire, G. (1997). *Calder* (Great modern masters series). New York: Abrams.

Marter, J. (1997). *Alexander Calder.* Cambridge: Cambridge University Press.

Schaefer, A. (2003). *Alexander Calder* (The life and works of . . . series). Chicago: Heinemann Library.

Related Instructional Resources

Aigner-Clark, J. (2004). *Baby Einstein: The ABCs of art.* New York: Hyperion Books for Children.

Dickins, R. (2005). *The children's book of art: An introduction to famous paintings.* London: Usborne.

Evans, J., & Skelton, T. (2001). *How to teach art to children.* Monterey, CA: Evan-Moor.

Kohl, M., & Solga, K. (1996). *Discovering great artists: Hands-on art for children in the styles of the great masters.* Bellingham, WA: Bright Ring.

Krull, K. (1995). *Lives of the artists: Masterpieces, messes.* San Diego: Harcourt Brace.

Muller, D., & Ward, R. (2007). Art and algebra? Middle school students discover algebra in Calder mobiles. *Mathematics in School, 36*(3), 15–21.

Renshaw, A., & Ruggi, G. (2005). *The art book for children.* New York: Phaidon Press.

Scieszka, J., & Smith, L. (2005). *Seen art?* New York: Viking Press.

Ward, R. & Muller, D. (September 2006). Algebra and art. *Mathematics Teaching, 198,* 22–26.

Related Websites

Alexander Calder
http://www.calder.org/
http://www.artcyclopedia.com/artists/calder_alexander.html
http://www.pbs.org/wnet/americanmasters/database/calder_a.html

History of Art Timeline
http://www.printsforpleasure.com/historyofart.htm

Interactive Calder Mobile
http://www.nga.gov/kids/zone/mobile.htm

Major Modern and Contemporary Visual Artists
http://www.the-artists.org/index.cfm

Study Art
http://www.sanford-artedventures.com/study/study.html

Teach Art
http://www.sanford-artedventures.com/teach/teach.html

Alexander Calder Worksheet

Name(s): ____

1. Record the weight of the pennies on both the left- and right-hand side of the lever. Also, record the distances from the fulcrum (balancing point) to the center of the stack of pennies on both sides.

Weight of Pennies on Left Side of Lever (W_L)	OPERATOR	Distance from fulcrum (D_L)		Weight of Pennies on Right Side of Lever (W_R)	OPERATOR	Distance from fulcrum (D_R)
			=			
			=			
			=			
			=			
			=			

2. What operation (addition, subtraction, multiplication, or division) must you perform on the weight and distance on the left-hand side of the lever to equal the same operation performed on the weight and distance on the right-hand side of the lever?

3. Express the Law of Levers as an algebraic equation.

Pablo Picasso (Famous Artists Series) (1994)

by Anthony Mason

Aladdin Books, ISBN #0812091752

Pablo Picasso: Breaking All the Rules (Smart about Art Series) (2002)

by True Kelley

Grosset & Dunlap, ISBN #0448428628

Overview of Books: In both of these biographies, learn about the life and artwork of the famous Spanish twentieth-century artist, Pablo Picasso, who, along with Georges Braque, cofounded an artistic period known as cubism.

Mathematical Concepts and Skills: Platonic Solids, two- and three-dimensional shapes, spatial reasoning, nets, faces, edges, vertices

Visual Arts Concepts and Skills: line, shape, form, space, creating art (cubist artwork), art history (life and artwork of Pablo Picasso)

Overview of Activities: Students explore the uniqueness of and then create nets of the five Platonic Solids (tetrahedron, cube, octahedron, dodecahedron, and icosahedron) and explore how two-dimensional shapes comprise three-dimensional solids. Students also discover the algebraic relationship among faces, edges, and vertices. Students then learn about the life and art of Pablo Picasso and the cubism art period and create their own piece of artwork in the spirit of cubism.

National Mathematics Standards (2000): Students in grades 6–8 should "precisely describe, classify, and understand relationships among types of two- and three-dimensional objects using their defining properties" (p. 397). Students should "use geometric models to represent and explain numerical and algebraic relationships" (p. 397). Students in grades 6–8 should also "recognize and apply geometric ideas and relationships in areas outside of the mathematics classroom, such as art, science, and everyday life" (p. 397) (Geometry Standard). Students in grades 6–8 should "represent, analyze, and generalize a variety of patterns with tables, graphs, words, and, when possible, symbolic rules" (p. 395) (Algebra Standard).

National Visual Arts Standards (1994): Students in grades 5–8 should use knowledge of structures and functions (Content Standard 2) and "employ organizational structures and analyze what makes them effective or not effective in the communication of ideas" (p. 50) (Achievement Standard b); choose and evaluate a range of subject matter, symbols, and ideas (Content Standard 3) and "integrate visual, spatial, and temporal concepts with content to communicate intended meaning in their artworks" (p. 50) (Achievement Standard a); reflect upon and assess the characteristics and merits of their work and the work of others (Content Standard 5) and "describe and compare a variety of individual responses to their own artworks and to artworks from various eras and cultures" (p. 51) (Achievement Standard c); make connections between visual arts and other disciplines (Content Standard 6) and "describe ways in which the principles and subject matter of other disciplines taught in the school are interrelated to the visual arts" (p. 51) (Achievement Standard b).

Materials: three-dimensional models of the five Platonic Solids (or access to interactive images of the Platonic Solids), paper, ruler, scissors, tape, images of Picasso's cubist paintings, rulers, scissors, gluesticks, 9" × 12" posterboard, magazines

Description of Activities:

1. Students view online images of the Platonic Solids using the Virtual Manipulatives—Platonic Solids website listed below, whereby students can rotate the solid and see all of its faces (or students can view a set of Platonic Solids created prior to class out of paper or plastic models). Begin a discussion about what students notice about the solids. Are they two- or three-dimensional? What shapes comprise each solid? Identify and define what are the faces, edges, and vertices. What makes this collection of these shapes unique? (Each solid is comprised of only one type of polygon.)
2. Using paper, ruler, scissors, and tape, challenge students to work collaboratively to create a net of the tetrahedron, cube, and octahedron. In doing so, students should notice that each particular Platonic Solid is comprised of only one type of polygon. For example, the tetrahedron and octahedron are comprised solely of triangles; the cube is comprised solely of squares. What sole polygon do students see in the dodecahedron and icosahedron? Students should fold their nets to verify that they did indeed create the intended three-dimensional shape.

3. Using their folded nets, students count and record in a table the number of faces, edges, and vertices in the tetrahedron, cube, and octahedron. Challenge students to determine the algebraic relationship among faces (F), edges (E), and vertices (V). Students should find that $F + V = E - 2$. Using the Virtual Manipulatives—Platonic Solids website listed below, students can verify if the formula holds true for the dodecahedron and icosahedron by pointing and clicking on each face, edge, and vertex.
4. Share excerpts from *Picasso* (Famous Artists Series) and *Pablo Picasso: Breaking All the Rules* (Smart about Art Series), which detail the life and works of Pablo Picasso, the famous twentieth-century Spanish artist. Spend time discussing cubism, a style of painting developed by Picasso and Georges Braque, in which the subject is captured from several sides or perspectives at once. The name *cubism* evolved from an insult by the French artist, Henri Matisse, who criticized cubism for being too mathematical and unemotional and being comprised of nothing but little cubes (one of the Platonic Solids). Share with students several of Picasso's cubist paintings, including *Three Musicians, Woman with Pears, Girl with Black Hair and Yellow Hat,* etc. Facilitate a discussion about what features students find appealing (or unappealing) about Picasso's cubist works.
5. Students create a piece of artwork in the spirit of Picasso's cubist style. Students cut out facial features of different individuals from a magazine (e.g., eyes, nose, mouth, chin, hair, etc.) into squares and rectangles of differing sizes. On 9" × 12" posterboard, students recreate a Picasso-like face using the individually shaped pieces.
6. Students discuss whose cubist artwork best captures the spirit of cubism.

Assessment:

- Did students recognize the Platonic Solids as a class of three-dimensional shapes, each of which is comprised of only one type of polygon?
- Did students correctly identify, count, and record the number of faces, edges, and vertices in the Platonic Solids?
- Did students discover the algebraic relationship among faces, edges, and vertices?
- Did students create a Picasso-like portrait and discuss whose artwork best captures the spirit of cubism?

Activity Extensions:

- Students explore other works of Picasso, including his blue period, rose period, etc.
- Students explore and then create their own artwork in the spirit of synthetic cubism, in which they glue objects and pieces of material to a canvas.
- Explore the work of other cubist artists including Georges Braque, Juan Gris, and others.

Cross-Curricular Connections:

Science

- Explore and discuss the Platonic Solids' associations with the elements in the universe.

Social Studies

- Explore and then share findings of a biographical piece of another cubist artist (e.g., Georges Braque, Juan Gris, etc.).

Related Literature

Antoine, V. (1994). *Picasso: A day in his studio* (Art for children series). New York: Chelsea House.

Bargallo, E. (2006). *My name is Picasso.* Hauppauge, NY: Barron's Educational Series.

Kelley, T. (2002). *Pablo Picasso: Breaking all the rules* (Smart about art series). New York: Grosset & Dunlap.

Lowery, L. (2006). *Pablo Picasso.* Minneapolis: Lerner.

Maltbie, P. (2005). *Picasso and Minou.* Watertown, MA: Charlesbridge.

Mason, A. (1994). *Pablo Picasso* (Famous artists series). London: Aladdin Books.

Mines, C. (2006). *Cubism.* Farmington Hills, MI: Thomas Gale.

Scarborough, K. (2001). *Pablo Picasso* (Artists in their times series). New York: Scholastic.

Venezia, M. (1988). *Picasso* (Getting to know the world's greatest artists series). New York: Scholastic.

Related Instructional Resources

Aigner-Clark, J. (2004). *Baby Einstein: The ABCs of art.* New York: Hyperion Books for Children.

Dickins, R. (2005). *The children's book of art: An introduction to famous paintings.* London: Usborne.

Evans, J., & Skelton, T. (2001). *How to teach art to children.* Monterey, CA: Evan-Moor.

Kohl, M., & Solga, K. (1996). *Discovering great artists: Hands-on art for children in the styles of the great masters.* Bellingham, WA: Bright Ring.

Krull, K. (1995). *Lives of the artists: Masterpieces, messes.* San Diego: Harcourt Brace.

Renshaw, A., & Ruggi, G. (2005). *The art book for children.* New York: Phaidon Press.

Scieszka, J., & Smith, L. (2005). *Seen art?* New York: Viking Press.

Related Websites

Cubism

http://www.abstractart.20m.com/cubism.htm
http://pablo-picasso.paintings.name/
http://www.sanford-artedventures.com/study/g_cubism.html
http://www.artcyclopedia.com/history/cubism.html

History of Art Timeline

http://www.printsforpleasure.com/historyofart.htm

Major Modern and Contemporary Visual Artists

http://www.the-artists.org/index.cfm

Pablo Picasso

http://www.princetonol.com/groups/iad/lessons/middle/sp-picas.htm
http://www.artchive.com/artchive/P/picasso.html
http://www.artchive.com/artchive/P/picasso_protocubism.html
http://www.picasso.fr/anglais/

Platonic Solids

http://www.mathsisfun.com/platonic_solids.html
http://mathforum.org/alejandre/workshops/net.html
http://www.korthalsaltes.com/
http://www.kheper.net/topics/cosmology/solids.html

Study Art

http://www.sanford-artedventures.com/study/study.html

Teach Art

http://www.sanford-artedventures.com/teach/teach.html

Virtual Manipulatives—Platonic Solids

http://nlvm.usu.edu/en/nav/frames_asid_128_g_3_t_3.html?open=instructions

Tessellations: The History and Making of Symmetrical Designs (2001)

by Pam Stephens

Crystal, ISBN #1562902431

M. C. Escher: The Graphic Work (2004)

by M. C. Escher

Taschen, ISBN #9707182563

Overview of Books: In *Tessellations: The History and Making of Symmetrical Designs,* follow your tour guide, Paul E. Gon, through a history of tessellations and the tesselated artwork of the twentieth-century Dutch artist, M. C. Escher, while learning the math and art of creating tessellations. Then, in *M. C. Escher: The Graphic Work,* explore the amazing visual artwork of M. C. Escher.

Mathematical Concepts and Skills: tessellations, patterns, transformations (translations, rotations, reflections, glide reflections)

Visual Arts Concepts and Skills: line, shape, creating art (tessellations), art history (life and artwork of M. C. Escher)

Overview of Activities: Students discover, explore, and create tessellations, which are unique patterns created by performing one or more transformation, such as a slide (translation), flip (reflection), turn (rotation), or glide reflection. Students also learn facts about the life and works of the Dutch artist M. C. Escher, famous for his tessellated artwork.

National Mathematics Standards (2000): Students in grades 6–8 should "describe sizes, positions, and orientations of shapes under informal transformations such as flips, turns, and slides" (p. 397). They should also "recognize and apply geometric ideas and relationships in areas outside of the mathematics classroom, such as art, science, and everyday life" (p. 397) (Geometry Standard).

National Visual Arts Standards (1994): Students in grades 5–8 should use knowledge of structures and functions (Content Standard 2) and "generalize about the effects of visual structures and functions and reflect upon these effects in their own work" (p. 50) (Achievement Standard a); make connections between visual arts and other disciplines (Content Standard 6) and "describe ways in which the principles and subject matter of other disciplines taught in the school are interrelated to the visual arts" (p. 51) (Achievement Standard b).

Materials: 4" × 6" index cards precut into 2" × 2" squares (approximately four for each student), paper, scissors, tape, glue sticks, construction paper, images of M. C. Escher's tessellated artwork

Description of Activities:

1. Begin by reading page 14 in *Tessellations: The History and Making of Symmetrical Designs,* which briefly introduces the Dutch artist, M. C. Escher, famous for his work containing optical illusions and tessellations. Share with students some examples of Escher's tessellated artwork appearing in *M. C. Escher: The Graphic Work* or using the M. C. Escher websites listed below. Ask students what features they notice in Escher's work (e.g., repeating pattern of the same shape). Introduce the concept of a tessellation, which is a repeating pattern with no gaps and no overlaps created by performing one or more transformations. Share more examples of tessellations shown in the Dale Seymour books (1991, 1997) or by using the Tessellations websites listed below.
2. Follow the instructions appearing on pages 18–37 in *Tessellations: The History and Making of Symmetrical Designs* and, using index cards precut into 2" × 2" squares, guide students in the creation of a tessellation by performing such transformations including a translation (slide), reflection (flip), rotation (turn), and glide reflection.
3. Students work independently and create their own two-color tessellation by performing one or more transformations on a new 2" × 2" tile. Students trace this tile eight to ten times onto a piece of colored construction paper. Students glue their colored tiles onto a different colored piece of colored construction paper, leaving spaces in between each tile, creating a two-colored alternating tessellated pattern.
4. Students share their tessellations with the class, challenging them to determine what transformation or series of transformations was used to create the tessellated pattern.

Assessment:

- Did students correctly perform various transformations (translation, reflection, rotation, and glide rotation)?
- Did students accurately create their own tessellation?
- Did students identify what transformation (or series of transformations) was used by other students in their tessellated artwork?

Activity Extensions:

- Explore various Escher works and challenge students to determine what transformation or series of transformations was used.
- Challenge students to find real-life examples of tessellations (brick wall, checkerboard, honeycomb, kitchen tile, roofing tiles, etc.).
- Enjoy poetry about reflections including "Reflection" (Silverstein, 1981), "Mirror" (Lewis, 1998), or "egamI rorriM ruoY mA I" (Prelutsky, 1996).

Cross-Curricular Connections:

Science

- Find examples of tessellations that occur in nature (e.g., honeycombs, snake skin, armadillo shell, etc.).
- Investigate how tessellations are used in architecture to improve strength in design (e.g., locking bricks in ancient pyramids).

Social Studies

- Look for tessellations in the woven patterns of Native American rugs and quilts.
- Explore tessellations in Islamic architecture (e.g., mosaic tile).

Related Literature

Brumbeau, J. (2001). *The quiltmaker's gift*. New York: Scholastic.

Brumbeau, J. (2004). *The quiltmaker's journey*. New York: Orchard Books.

Escher, M. (2004). *M. C. Escher: The graphic work*. Hohenzollernring, Germany: Taschen.

Friedman, A. (1994). *A cloak for the dreamer*. New York: Scholastic.

Jonas, A. (1987). *Reflections*. New York: Greenwillow Books.

Lewis, J. (1998). *Doodle dandies: Poems that take shape*. New York: Aladdin.

Prelutsky, J. (1996). *A pizza the size of the sun*. New York: Scholastic.

Roza, G. (2005). *An optical artist: Exploring patterns and symmetry*. New York: PowerKids Press.

Silverstein, S. (1981). *A light in the attic*. New York: HarperCollins.

Stephens, P. (2001). *Tessellations: The history and making of symmetrical designs*. Aspen, CO: Crystal.

Related Instructional Resources

Dale Seymour. (1991). *Tessellation winners: Escher-like original student art, the first contest*. Palo Alto, CA: Dale Seymour.

Dale Seymour. (1997). *Tessellation teaching masters*. Palo Alto, CA: Dale Seymour.

Evans, J., & Skelton, T. (2001). *How to teach art to children*. Monterey, CA: Evan-Moor.

Kohl, M., & Solga, K. (1996). *Discovering great artists: Hands-on art for children in the styles of the great masters*. Bellingham, WA: Bright Ring.

Krull, K. (1995). *Lives of the artists: Masterpieces, messes*. San Diego: Harcourt Brace.

Renshaw, A., & Ruggi, G. (2005). *The art book for children*. New York: Phaidon Press.

Scieszka, J., & Smith, L. (2005). *Seen art?* New York: Viking Press.

Shattschneider, D. (2005). *M. C. Escher: Visions of symmetry*. Petaluma, CA: Pomegranate.

Shattschneider, D., & Walker, W. (2005). *M. C. Escher kaleidocycles*. New York: Abrams.

Related Websites

Iproject—Make Your Own Tessellation
http://www.iproject.com/escher/teaching/maketessel.html

History of Art Timeline
http://www.printsforpleasure.com/historyofart.htm

Major Modern and Contemporary Visual Artists
http://www.the-artists.org/index.cfm

M. C. Escher
http://www.mcescher.com/
http://www.artcyclopedia.com/artists/escher_mc.html

M. C. Escher and Math (the Droste Effect)
http://escherdroste.math.leidenuniv.nl/

Study Art
http://www.sanford-artedventures.com/study/study.html

Teach Art
http://www.sanford-artedventures.com/teach/teach.html

Tessellations
http://britton.disted.camosun.bc.ca/jbsymteslk.htm
http://mathforum.org/sum95/suzanne/tess.intro.html
http://www.tessellations.org/
http://mathforum.org/sum95/suzanne/tess.intro.html
http://library.thinkquest.org/16661/escher/tessellations.1.html

Virtual Manipulatives Library—Reflections
http://nlvm.usu.edu/en/nav/frames_asid_297_g_2_t_3.html?open=activities

Virtual Manipulatives Library—Rotations
http://nlvm.usu.edu/en/nav/frames_asid_299_g_2_t_3.html?open=activities

Virtual Manipulatives Library—Tessellations
http://nlvm.usu.edu/en/nav/frames_asid_163_g_2_t_3.html?open=activities

Virtual Manipulatives Library—Translations
http://nlvm.usu.edu/en/nav/frames_asid_301_g_2_t_3.html?open=activities

Activities Featuring Measurement

Mondrian (Great Modern Masters Series) (1997)

by Jose Maria Faerna

Abrams, ISBN #0810946874

Overview of Book: Readers learn about the life and artwork of the twentieth-century Dutch abstract painter, Piet Mondrian, whose work primarily consisted of primary-colored rectangles formed by vertical and horizontal lines.

Mathematical Concepts and Skills: parallel lines, perpendicular lines, intersecting lines, transversal, angles (vertical, supplementary, complementary, alternate interior, alternate exterior, corresponding, consecutive interior), angle measurement, angle relationships

Visual Arts Concepts and Skills: line, shape, color, creating art (abstract art), art history (life and artwork of Piet Mondrian)

Overview of Activities: By viewing two parallel lines cut by a transversal, students measure angles and learn the relationships between and among the various types of angles including vertical angles, corresponding angles, alternate interiors angles, alternate exterior angles, etc. Students then explore how Mondrian used vertical and horizontal lines to form his *Composition* series. Students create a piece of artwork in the spirit of Mondrian by drawing a series of perpendicular lines and then coloring the resulting rectangular regions using the primary colors.

National Mathematics Standards (2000): Students in grades 6–8 should "understand, select, and use units of appropriate size and type to measure angles" (p. 399). Students should "select and apply techniques and tools to accurately find angle measures" (p. 399) (Measurement Standard). Students in grades 6–8 should also "understand relationships among the angles, side lengths, perimeters, areas, and volumes of similar objects" and "recognize and apply geometric ideas and relationships in areas outside of the mathematics classroom, such as art, science, and everyday life" (p. 397) (Geometry Standard).

National Visual Arts Standards (1994): Students in grades 5–8 should use knowledge of structures and functions (Content Standard 2) and "employ organizational structures and analyze what makes them effective or not effective in the communication of ideas" (p. 50) (Achievement Standard b); choose and evaluate a range of subject matter, symbols, and ideas (Content Standard 3) and "integrate visual, spatial, and temporal concepts with content to communicate intended meaning in their artworks" (p. 50) (Achievement Standard a); reflect upon and assess the characteristics and merits of their work and the work of others (Content Standard 5) and "describe and compare a variety of individual responses to their own artworks and to artworks from various eras and cultures" (p. 51) (Achievement Standard c); make connections between visual arts and other disciplines (Content Standard 6) and "describe ways in which the principles and subject matter of other disciplines taught in the school are interrelated to the visual arts" (p. 51) (Achievement Standard b).

Materials: paper, rulers, protractors, 9" × 12" posterboard, black construction paper, scissors, crayons, glue sticks

Description of Activities:

1. Students draw two intersecting lines (but not perpendicular). Students view the four angles created and estimate their angle measure. Students use protractors to measure each angle. What do students notice?
2. Define vertical angles (angles that are formed by two intersecting lines that are opposite each other and have congruent measure). Define supplementary angles (two angles whose sum is 180 degrees). Students use their sketches and angle measurements to identify which angles are vertical and which angles are supplementary. Define complementary angles (two angles that sum to 90 degrees). Does their sketch contain any complementary angles?
3. Using a ruler, students draw two horizontal lines situated about 4 inches apart. Students draw a diagonal line that intersects both of the horizontal lines. Define this line as a transversal (a line that cuts across a set of lines). Students will notice that eight angles are created. Students label each angle and then use their protractors to measure each angle. Which angles are vertical? Supplementary?

4. Students identify and record other angles of equal measure. Assist students in naming and locating the alternate interior angles (two interior angles that lie on different parallel lines and on opposite sides of a transversal), the alternate exterior angles (two exterior angles on opposite sides of a transversal that lie on different parallel lines), and the corresponding angles (two angles, one of which is an interior angle, the other an exterior angle, located on the same side of the transversal and are in corresponding positions). Students also identify and locate the consecutive interior angles (two interior angles lying on the same side of the transversal cutting across two parallel lines and which are supplementary).
5. Show samples of the artwork of the twentieth-century Dutch abstract artist, Piet Mondrian, using the book *Mondrian* (Great Modern Masters Series) or one of the Piet Mondrian websites listed below. Mondrian grew up in Holland and created abstract art to express his thoughts and feelings. Encourage students to notice the vertical and horizontal lines traversing Mondrian's paintings that result in a series of squares and rectangles. Ask students why they think Mondrian is labeled an abstract painter. What do students think is the meaning or message behind his artwork? What do students find appealing (or unappealing) about Mondrian's works? Facilitate a discussion.
6. Students create a Mondrian-like painting. Students cut black construction paper into thin strips (about 1 cm in width). Students purposefully place and then glue the black strips onto a 9" × 12" piece of posterboard, placing them both vertically and horizontally, creating lines that are perpendicular. This will result in the creation of different sized square and rectangular regions. Students should take care in placing the black strips, evaluating how their placement affects the overall visual appearance of their artwork. Using red, yellow, and blue crayons, students color in the regions.
7. Hang the students' artwork and let students compare and contrast their work, articulating their appeal.

Assessment:

- Did students accurately estimate, measure, and label the various types of angles?
- Did students participate in discussion about the appeal of Mondrian's work and its meaning?
- Did students create a Mondrian-like painting?
- Did students respectfully critique each other's Mondrian-like works?

Activity Extensions:

- Students find examples of angles (vertical, supplementary, etc.) in the classroom or at home.
- Explore the concept of area using a Mondrian painting. Students compute and verify that the sum of the areas of all of the rectangular regions in a painting equals the area of the entire painting.
- Visit the Mondrimat website listed below and let students experiment with color and shape by creating their own online artwork in the spirit of Mondrian.

Cross-Curricular Connections:

Science

- Since primary colors are important to Mondrian's work, investigate and discuss chromatics, the science of color through paint mixing, prisms, and rainbows.
- Identify angles occurring naturally in the environment.

Social Studies

- Explore and then share findings of a biographical piece of another abstract artist (e.g., Willem de Kooning, Jackson Pollock, etc.).

Related Literature

Faerna, J. (1997). *Mondrian* (Great modern masters series). New York: Abrams.

Grange Books. (2005). *Mondrian*. London: Grange Books.

Ljungkvist, L. (2006). *Follow the line*. New York: Penguin Group.

Milner, J. (1995). *Mondrian*. New York: Phaidon Press.

Related Instructional Resources

Aigner-Clark, J. (2004). *Baby Einstein: The ABCs of art*. New York: Hyperion Books for Children.

Dickins, R. (2005). *The children's book of art: An introduction to famous paintings*. London: Usborne.

Evans, J., & Skelton, T. (2001). *How to teach art to children*. Monterey, CA: Evan-Moor.

Fauchereau, S. (1994). *Mondrian*. New York: Rizzoli.

Kohl, M., & Solga, K. (1996). *Discovering great artists: Hands-on art for children in the styles of the great masters*. Bellingham, WA: Bright Ring.

Krull, K. (1995). *Lives of the artists: Masterpieces, messes*. San Diego: Harcourt Brace.

Renshaw, A., & Ruggi, G. (2005). *The art book for children*. New York: Phaidon Press.

Scieszka, J., & Smith, L. (2005). *Seen art?* New York: Viking Press.

Related Websites

Abstract Art
http://www.sanford-artedventures.com/study/g_abstract.html

History of Art Timeline
http://www.printsforpleasure.com/historyofart.htm

Major Modern and Contemporary Visual Artists
http://www.the-artists.org/index.cfm

Measuring Angles
http://www.kidport.com/Grade5/Math/MeasureGeo/MeasuringAngles.htm

Mondrimat
http://www.stephen.com/mondrimat/

Piet Mondrian
http://www.artcyclopedia.com/artists/mondrian_piet.html
http://www.storyboardtoys.com/gallery/Piet-Mondrian.htm
http://www.mossfoundation.org/page.php?id=81
http://www.ibiblio.org/wm/paint/auth/mondrian/

Study Art
http://www.sanford-artedventures.com/study/study.html

Teach Art
http://www.sanford-artedventures.com/teach/teach.html

Paul Cezanne (Famous Artists Series) (1994)

by Anthony Mason

Barron's Educational Series, ISBN #0812012933

Paul Cezanne: A Painter's Journey (2006)

by Robert Burleigh

Abrams, ISBN #0810957841

Overview of Books: Enjoy both of these biographies of the nineteenth-century French post-impressionist artist, Paul Cezanne, who is often described as the Father of Modern Art, although his art was misunderstood and discredited by the public during most of his life. Cezanne's later works, in which he experimented with geometric simplification and optical phenomena, served as a catalyst for the period of cubism.

Mathematical Concepts and Skills: three-dimensional shapes, cones, cylinders, pyramids, prisms, volume

Visual Arts Concepts and Skills: line, shape, color, value, form, space, creating art (still life), art history (life and artwork of Paul Cezanne)

Overview of Activities: Students explore and learn about the life and artwork of Paul Cezanne, who considered the sphere, cone, and cylinder to be the basic forms. Students then explore the relationship between the volume of a cone and cylinder and the relationship between the volume of a pyramid and prism. Students create a piece of still life in the spirit of Cezanne.

National Mathematics Standards (2000): Students in grades 6–8 should "develop strategies to determine the surface area and volume of selected prisms, pyramids, and cylinders" (p. 399) (Measurement Standard).

National Visual Arts Standards (1994): Students in grades 5–8 should use knowledge of structures and functions (Content Standard 2) and "employ organizational structures and analyze what makes them effective or not effective in the communication of ideas" (p. 50) (Achievement Standard b); reflect upon and assess the characteristics and merits of their work and the work of others (Content Standard 5) and "describe and compare a variety of individual responses to their own artworks and to artworks from various eras and cultures" (p. 51) (Achievement Standard c); make connections between visual arts and other disciplines (Content Standard 6) and "describe ways in which the principles and subject matter of other disciplines taught in the school are interrelated to the visual arts" (p. 51) (Achievement Standard b).

Materials: images of Cezanne's works (e.g., *Apples and Oranges, Millstone and Cistern under Trees, Still Life with Peppermint Bottle, Kitchen Still Life, Mont Sainte-Victoire Seen from Les Lauves, House of Pere Lacroix,* etc.); paper plates; paper models of cylinders, cones, square pyramids, and cubes (made prior to class); popcorn kernels; small cup or scoop; oil paints; large white paper

Description of Activities:

1. Share with students select passages about the life and artwork of Paul Cezanne from *Cezanne* (Famous Artists Series) and *Paul Cezanne: A Painter's Journey*. Cezanne was a French post-impressionist painter whose work bridged late nineteenth century impressionism and early twentieth century cubism. His earlier works were expressionistic, and then, in the early 1870s, he experimented with impressionism (e.g., *Le Chateau de Medan*). After that, Cezanne delved into classicism (e.g., *Still Life with Plaster Cupid*) and, toward the end of his life, he reduced architecture and figures to geometric forms, thus paving the way for cubism (e.g., *Bend in the Road*).

2. Cezanne considered the sphere, cone, and cylinder to be basic forms. Share with students several examples of Cezanne's works, encouraging them to locate these shapes (e.g., *Apples and Oranges, Millstone and Cistern under Trees, Still Life with Peppermint Bottle, Kitchen Still Life,* etc.).

3. Place students into small groups. Using the Nets of Cones, Cylinders, Prisms, and Pyramids websites listed below, provide the groups with a cylinder and a cone (created prior to class). Encourage students to notice the similarities between the cylinder and cone (both have the same height and same sized circular base). Ask students to visually estimate the volume of the cone relative to the cylinder. That is, if a student were to fill up the cylinder, how many level cones would fit inside the cylinder? Let students share their estimates and justify their reasoning. (Students might guess that two cones fit inside a cylinder, although three cones actually do.)

4. Distribute a bag of popcorn kernels (or rice), a small cup or scoop, and a paper plate to the groups. Using the scoop, students fill up the cone with the popcorn kernels until it is level, and then carefully pour the coneful into the cylinder (which is held by one student on a paper plate in case of spilling). Students should notice that the cylinder is not half full, as they may have anticipated. Students fill the cone again with popcorn kernels and then carefully pour the coneful into the cylinder. Students will see that one more coneful is needed. Thus, three cones fill the cylinder.
5. Challenge students to algebraically express the relationship between the volume of a cone and the volume of a cylinder (with the same height and same sized base). The volume of a cone is one-third that of the volume of a cylinder. That is, volume of cone = $1/3 \times$ (volume of a cylinder), or volume of cone = $1/3\ \pi r^2 h$ (where πr^2 is the area of the cylinder's base and h is its height). The factor of one-third appears in the volume formula because three conefuls of popcorn fit inside the cylinder.
6. Distribute to students a paper model of one square pyramid and one cube (both constructed using one of the Nets websites listed below and created prior to class). Challenge students to visually determine the relationship between the square pyramid and the cube. Distribute two more square pyramids to the groups and let students experiment with positioning the pyramids to form a cube. Students algebraically express the volume of a square pyramid in terms of a cube. The volume of a square pyramid is one-third that of a cube (of the same dimensions). Further, the volume of any pyramid is one-third that of a prism with the same sized base and height.
7. Students work in two large groups. Distribute oil paints and large white paper to students. Students bring to class an apple or orange. Students create a still life scene similar to Cezanne's *Apples and Oranges,* using their fruit and other props (e.g., bowl, pitcher, cloth, etc.). Students paint a still life in the spirit of Cezanne, focusing on using cones, cylinders, and spheres as the basic shapes, as did Cezanne. Students share and discuss their artwork with the class.

Assessment:

- Did students notice cylinders, cones, and spheres in Cezanne's works?
- Did students correctly express the volume of a cone and pyramid?
- Did students create a still life painting in the spirit of Cezanne?

Activity Extensions:

- Students explore how to compute surface area of cylinders, cones, pyramids, and prisms.
- Students locate real-life examples of cylinders, cones, pyramids, and prisms.
- Students explore the work of other artists who used shapes in their paintings (e.g., Joan Miro, Henri Matisse, Wassily Kandinsky, etc.).

Cross-Curricular Connections:

Science

- Explore and discuss the engineering knowledge and skills required to build pyramids.

Social Studies

- Explore and then share findings of a biographical piece of another artist influenced by Cezanne (e.g., Pablo Picasso, Georges Braque, Juan Gris, etc.).

Related Literature

Burleigh, R. (2006). *Paul Cezanne: A painter's journey.* New York: Abrams.

Connelly, S. (2006). *Paul Cezanne* (The life and work of . . . series). Chicago: Heinemann Library.

Harris, N. (2003). *Paul Cezanne* (Artists in their time series). New York: Scholastic.

Mason, A. (1994). *Paul Cezanne* (Famous artists series). Hauppauge, NY: Barron's Educational Series.

Sellier, M. (1996). *Cezanne from A to Z.* New York: Bedrick.

Venezia, M. (1988). *Paul Cezanne* (Getting to know the world's greatest artists series). Chicago: Children's Press.

Related Instructional Resources

Aigner-Clark, J. (2004). *Baby Einstein: The ABCs of art.* New York: Hyperion Books for Children.

Boutan, M. (1995). *Cezanne.* San Francisco: Raincoast Books.

Cros, P. (2006). *Cezanne.* New York: Vilo.

Dean, C. (1993). *Cezanne.* New York: Phaidon Press.

Dickins, R. (2005). *The children's book of art: An introduction to famous paintings.* London: Usborne.

Evans, J., & Skelton, T. (2001). *How to teach art to children.* Monterey, CA: Evan-Moor.

Kohl, M., & Solga, K. (1996). *Discovering great artists: Hands-on art for children in the styles of the great masters.* Bellingham, WA: Bright Ring.

Krull, K. (1995). *Lives of the artists: Masterpieces, messes.* San Diego: Harcourt Brace.

Renshaw, A., & Ruggi, G. (2005). *The art book for children.* New York: Phaidon Press.

Scieszka, J., & Smith, L. (2005). *Seen art?* New York: Viking Press.

Related Websites

History of Art Timeline
http://www.printsforpleasure.com/historyofart.htm

Interactive Volume/Fill and Pour Activity
http://nlvm.usu.edu/en/nav/frames_asid_275_g_3_t_4.html

Major Modern and Contemporary Visual Artists
http://www.the-artists.org/index.cfm

Net (and Graphic) of Pyramid and Cube
http://www.korthalsaltes.com/three_pyramids_in_a_cube.htm
http://www.korthalsaltes.com/pdf/three_pyramids_in_one_cube.pdf

Nets of Cones, Cylinders, Prisms, and Pyramids
http://www.korthalsaltes.com/index_hoofd.html
http://lrt.ednet.ns.ca/PD/BLM/table_of_contents.htm
http://schools.ednet.ns.ca/avrsb/732/hainstoc/briefcase/techintegpage_math_blacklinemasters.html

Paul Cezanne
http://gardenofpraise.com/art47.htm
http://www.ibiblio.org/wm/paint/auth/cezanne/
http://www.ibiblio.org/wm/paint/auth/cezanne/style.html
http://www.artcyclopedia.com/artists/cezanne_paul.html

Study Art
http://www.sanford-artedventures.com/study/study.html

Teach Art
http://www.sanford-artedventures.com/teach/teach.html

Miro: Earth and Sky (Art for Children Series) (1993)

by Claire-Helene Blanquet

Chelsea House, ISBN #0791028135

Joan Miro (Famous Artists Series) (1995)

by Nicholas Ross

Barron's Educational Series, ISBN #0812094271

Overview of Books: In *Miro: Earth and Sky* (Art for Children Series), enjoy biographical facts and direct quotes from Joan Miro interwoven into a story about a young girl whose teacher points out that her sleeping bag is decorated and patterned after a Miro painting. Thus, the class embarks on a journey into the life and works of the Spanish surrealist painter and sculptor. In *Joan Miro* (Famous Artists Series), enjoy a biography as well as images of the works of Joan Miro, who was considered to be one of the most versatile masters of twentieth-century art.

Mathematical Concepts and Skills: probability, random events, likelihood, fair chance, theoretical probability, experimental probability, Law of Large Numbers, prediction

Visual Arts Concepts and Skills: line, shape, color, form, space, creating art (surrealist art), art history (life and artwork of Joan Miro)

Overview of Activities: Students explore theoretical probability, experimental probability, and the Law of Large Numbers by carrying out an experiment involving dice. Students then learn about the life and artwork of Joan Miro, the Spanish surrealist famous for creating his works by drawing by chance, a technique of random, imaginative drawing. Students create a piece of art in the spirit of Miro.

National Mathematics Standards (2000): Students in grades 6–8 should "compute probabilities for simple compound events, using such methods as organized lists, tree diagrams, and area models" (p. 401). Students should also "understand and use appropriate terminology to describe complementary and mutually exclusive events" (p. 401). Students should "use proportionality and a basic understanding of probability to make and test conjectures about the results of experiments and simulations" (p. 401) (Data Analysis and Probability Standard).

National Visual Arts Standards (1994): Students in grades 5–8 should choose and evaluate a range of subject matter, symbols, and ideas (Content Standard 3) and "use subjects, themes, and symbols that demonstrate knowledge of contexts, values, and aesthetics that communicate intended meaning in artworks" (p. 50) (Achievement Standard b); reflect upon and assess the characteristics and merits of their work and the work of others (Content Standard 5) and "describe and compare a variety of individual responses to their own artworks and to artworks from various eras and cultures" (p. 51) (Achievement Standard c); make connections between visual arts and other disciplines (Content Standard 6) and "describe ways in which the principles and subject matter of other disciplines taught in the school are interrelated to the visual arts" (p. 51) (Achievement Standard b).

Materials: six-sided dice, images of Miro's works (e.g., *Harlequin's Carnival, Dutch Interior I, The Smile of Flaming Wings, Woman Encircled by the Flight of a Bird, For Emili Fernandez Miro, Woman and Bird in the Night, Wall of the Moon, Wall of the Sun*, etc.), black pen, oil paints, large white paper

Description of Activities:

1. Begin a discussion in which students articulate their working definitions of probability (a measure of the likelihood of an event occurring), fair events (fair events are those which have an equal likelihood of occurring), a random event (an event that is allowed to happen without any attempt to control the event's outcome), and other relevant probability terms. Challenge students to give an example of a random event, a fair event, and so on.

2. Students work in pairs and are given a six-sided die. Students view the die and discuss whether each number has an equal likelihood of being rolled. Introduce, define, and discuss the terms *theoretical probability* (the chance that an event would happen under perfect circumstances) and *experimental* (or empirical) *probability* (the ratio of the number of times an event occurred to the total number of times the event was carried out). Thus, the theoretical probability of rolling a four is one out of six (or $\frac{1}{6}$, since there is one chance in six that a four will appear). However, after rolling a die six times, if a four appeared two times, the experimental, or empirical, probability is two out of six (or $\frac{2}{6}$ or $\frac{1}{3}$).

3. Pairs of students roll their die thirty to fifty times and tally and record the resulting outcomes in a table. Students then compute and record the experimental probability for each number. Did the experimental probability for each number match its theoretical probability of $\frac{1}{6}$? Why or why not? Was this experiment fair?

4. On the board, sum all the pairs' data for each number one through six. Do the experimental probabilities more closely match the theoretical probabilities? Generate a discussion and introduce the Law of Large Numbers (also called Bernoulli's Law), which states that if an experiment is carried out a large number of times, the experimental probability of the event occurring tends toward the theoretical probability of the event occurring.
5. Share with students select passages about the life and artwork of Joan Miro from *Miro: Earth and Sky* and *Joan Miro*. Many of Miro's works (e.g., *Harlequin's Carnival, Dutch Interior I, The Smile of Flaming Wings*, etc.) have a whimsical and humorous quality, containing images of playfully distorted animal forms, twisted shapes, and odd geometric constructions. Miro has been described as drawing by chance, a technique of random, imaginative drawing, evidenced in his *Constellations* series. The talented artist also sculpted, worked in ceramics, and produced many mosaics and murals. Two of his most notable sculptures are two large ceramic murals entitled *Wall of the Moon* and *Wall of the Sun*, created for the UNESCO building in Paris. Share with students a variety of Miro's works. Facilitate a discussion about what features students find appealing (or unappealing) about his works.
6. Implement the activity described on page 19 in *Joan Miro* where students create a Miro work by using his technique called drawing by chance. Provide students with a thick black marker, oil paints (or another coloring medium), and large white paper. Using their marker, students close their eyes and make bold, random marks on their paper. Students open their eyes and, using their imagination, look for and complete figures and shapes. Fill in the shapes using colored oil paints. Students then think of a creative title for the painting, as did Miro.
7. Students share their artwork with the class. Students discuss whose artwork is most visually appealing and why, and whose title was most creative.

Assessment:

- Did students provide working definitions of various probability terms?
- Did students accurately compute the experimental probabilities in their die rolling experiment?
- Did students create a piece of artwork in the spirit of Miro?
- Did students aptly name their Miro work and discuss the appeal of other students' artwork?

Activity Extensions:

- Explore the bell curve by rolling two dice fifty times, computing the sums, and then tallying the number of times each sum occurs.
- Carry out other probability experiments using dice with eight or more sides or by visiting the Virtual Manipulatives—Coin Tossing or Spinners websites listed below.
- Visit the Random Drawing Tool website, an applet that simulates randomly drawing tickets from a box.
- Visit the Random Birthday Applet website, which randomly generates a number of dates representing birthdays and tabulates when two of the generated dates match.

Cross-Curricular Connections:

Science

- Miro's most famous series of paintings was his *Constellations* series. Explore and discuss stars and constellations.

Social Studies

- Research and discuss the geography and other demographics of the Catalonia region of Spain, where Joan Miro was born and lived most of his life.

Related Literature

Blanquet, C. (1993). *Miro: Earth and sky* (Art for children series). New York: Chelsea House.

Malet, R. (2003). *Joan Miro.* New York: Rizzoli.

Mink, J. (2000). *Miro.* Hohenzollernring, Germany: Taschen.

Ross, N. (1995). *Joan Miro* (Famous artists series). Hauppauge, NY: Barron's Educational Series.

Related Instructional Resources

Aigner-Clark, J. (2004). *Baby Einstein: The ABCs of art.* New York: Hyperion Books for Children.

Dickins, R. (2005). *The children's book of art: An introduction to famous paintings.* London: Usborne.

Evans, J., & Skelton, T. (2001). *How to teach art to children.* Monterey, CA: Evan-Moor.

Kohl, M., & Solga, K. (1996). *Discovering great artists: Hands-on art for children in the styles of the great masters.* Bellingham, WA: Bright Ring.

Krull, K. (1995). *Lives of the artists: Masterpieces, messes.* San Diego: Harcourt Brace.

Renshaw, A., & Ruggi, G. (2005). *The art book for children.* New York: Phaidon Press.

Scieszka, J., & Smith, L. (2005). *Seen art?* New York: Viking Press.

Related Websites

History of Art Timeline
http://www.printsforpleasure.com/historyofart.htm

Joan Miro
http://www.artnet.com/artist/675008/joan-miro.html
http://www.artcyclopedia.com/artists/miro_joan.html
http://www.artelino.com/articles/joan_miro.asp
http://www.enchantedlearning.com/artists/miro/
http://www.halter.net/gallery/miro.html

Major Modern and Contemporary Visual Artists
http://www.the-artists.org/index.cfm

Random Birthday Applet
http://www-stat.stanford.edu/~susan/surprise/Birthday.html

Random Drawing Tool
http://illuminations.nctm.org/ActivityDetail.aspx?id=67

Study Art
http://www.sanford-artedventures.com/study/study.html

Surrealist Period and Painters
http://www.sanford-artedventures.com/study/g_surrealism.html
http://www.artcyclopedia.com/history/surrealism.html

Teach Art
http://www.sanford-artedventures.com/teach/teach.html

Virtual Manipulatives—Coin Tossing
http://nlvm.usu.edu/en/nav/frames_asid_305_g_3_t_5.html

Virtual Manipulatives—Spinners
http://nlvm.usu.edu/en/nav/frames_asid_186_g_3_t_5.html?open=activities

Arp (1988)

by Serge Fauchereau

Rizzoli, ISBN #0847809749

Looking at Dada (2006)

by Sarah Ganz Blythe and Edward D. Powers

The Museum of Modern Art, ISBN #0870707051

Overview of Books: In *Arp*, discover the life and works of the twentieth-century French-German sculptor, writer, painter, and poet, Hans (Jean) Arp, one of the founders of the art movement known as dada. In *Looking at Dada*, learn about dada, a movement that emerged during World War I and that challenged artistic norms and social values. Thirty major artists (such as Jean Arp, Marcel Duchamp, Max Ernst, and others) and their works (including paintings, sculptures, printmaking, poetry, photography, collage, etc.) are featured in *Looking at Dada*.

Mathematical Concepts and Skills: probability, random events, Law of Large Numbers, prediction, pi

Visual Arts Concepts and Skills: line, shape, color, value, texture, space, creating art (chance collage), art history (life and artwork of Jean Arp and Dada)

Overview of Activities: Students predict and compute the probability of events by carrying out an experiment involving dropping a needle onto lined paper. Students then learn about the life and artwork of Jean Arp, a dadaist famous for creating his collages by chance. Students create a collage in the spirit of Arp.

National Mathematics Standards (2000): Students in grades 6–8 should "compute probabilities for simple compound events, using such methods as organized lists, tree diagrams, and area models" (p. 401). Students should "use proportionality and a basic understanding of probability to make and test conjectures about the results of experiments and simulations" (p. 401) (Data Analysis and Probability Standard).

National Visual Arts Standards (1994): Students in grades 5–8 should use knowledge of structures and functions (Content Standard 2) and "generalize about the effects of visual structures and functions and reflect upon these effects in their own work" (p. 50) (Achievement Standard a); choose and evaluate a range of subject matter, symbols, and ideas (Content Standard 3) and "use subjects, themes, and symbols that demonstrate knowledge of contexts, values, and aesthetics that communicate intended meaning in artworks" (p. 50) (Achievement Standard b); reflect upon and assess the characteristics and merits of their work and the work of others (Content Standard 5) and "describe and compare a variety of individual responses to their own artworks and to artworks from various eras and cultures" (p. 51) (Achievement Standard c); make connections between visual arts and other disciplines (Content Standard 6) and "describe ways in which the principles and subject matter of other disciplines taught in the school are interrelated to the visual arts" (p. 51) (Achievement Standard b).

Materials: sheets of paper with a series of vertical lines drawn on them with each line 1 inch apart (prepared prior to class), 1-inch long needles or pins, calculators, colored tissue paper or construction paper cut into 1-inch squares, 9" × 12" posterboard, glue sticks, image of Arp's *Collage with Squares Arranged According to the Laws of Chance*

Description of Activities:

1. Begin a discussion in which students articulate their working definitions of probability (a measure of the likelihood of an event occurring), fair events (fair events are those which have an equal likelihood of occurring), a random event (an event that is allowed to happen without any attempt to control the event's outcome), and other relevant probability terms. Challenge students to give an example of a random event, a fair event, the probability of an event occurring, and so on.

2. Distribute to pairs of students a sheet of paper with a series of vertical lines drawn on it with each line 1 inch apart (made prior to class) and a needle or pin 1 inch in length. Ask students to describe the possible resulting outcomes if they dropped the needle on the paper (either the needle crosses a line or it does not). Define a toss as *favorable* if the needle crosses a line and as *unfavorable* if the needle does not cross a line. Ask students to discuss and predict the probability of each of these events. Are these events fair?

3. Students carry out an experiment in which they drop a needle fifty times onto their paper, recording each time whether the needle crossed a line (favorable toss) or did not cross a line (unfavorable toss).
4. Students compute the ratio $\frac{2N}{C}$, where N represents the total number of times the needle was dropped (in this case, fifty), and C is the number of times the toss was favorable. Students will find that the ratio $\frac{2N}{C}$ approaches the value of pi (3.1415...).
5. To develop a closer approximation of pi, combine all of the pairs' data, resulting in a new and larger value for N (total number of times the needle was dropped) and C (the number of favorable tosses). Does this value come closer to approximating pi? Why or why not?
6. Generate a discussion and introduce the Law of Large Numbers (also called Bernoulli's Law), which states that if an experiment (in this case, the dropping of the needle on a sheet of paper) is carried out a large number of times, the experimental probability of the event occurring tends toward the theoretical probability of the event occurring.
7. Visit the Buffon's Needle websites listed below, which simulate this experiment and discover the proof of this fascinating outcome.
8. Read excerpts from *Looking at Dada* and introduce and discuss dada, which was a movement in the arts that began in Switzerland during World War I and peaked during 1916–1920. As a way to protest the war, dadaists created anti-art, protesting against the contemporary academic and cultured values of art. Dada was based on the principles of deliberate irrationality and the rejection of the laws of beauty and social organization. The works of the dadaists later influenced surrealist and pop artists.
9. Share with students select passages about the life and artwork of Jean Arp from *Looking at Dada* (pp. 27–32) as well as from Arp (pp. 7–29). Jean Arp was a French-German sculptor, writer, painter, and poet. A believer in the artistic laws of chance, he invented a new kind of collage technique in which he tore paper into squares and dropped them from above, watching them land on a piece of paper below. Despite not knowing where the torn squares would land, Arp still loved the outcome of this design by chance. Show students an image of Arp's *Collage with Squares Arranged According to the Laws of Chance,* appearing in *Looking at Dada* (p. 30) or using the Jean Arp's *Collage with Squares Arranged According to the Laws of Chance* website listed below. Facilitate a discussion about what features students find appealing (or unappealing) about this particular work. Share with students other

images of Arp's random collages appearing in *Arp* (e.g., *Random Collage*, p. 38; *Torn Paper* and *Second Torn Paper*, p. 45, etc.).

10. Students create a chance collage in the spirit of Arp by dropping pieces of colored construction paper or tissue paper cut into 1-inch squares onto a 9" × 12" piece of posterboard. (Students should spread their glue sticks onto the posterboard before dropping the torn pieces, allowing the square pieces to stick where they land.) Students share their artwork with the class. Students discuss whose artwork is most visually appealing and why.

Assessment:

- Did students provide working definitions of various probability terms?
- Did students accurately tally the outcomes of their needle dropping experiment?
- Did the students' experiment result in a close approximation of pi?
- Did students create a piece of artwork in the spirit of Arp?
- Did students discuss the appeal of Arp's works as well as the appeal of other students' Arp-like collages?

Activity Extensions:

- Further explore the Law of Large Numbers by using the Virtual Manipulatives—Coin Tossing website listed below.
- Carry out interactive experiments and explore the concepts of random events and probability using the Virtual Manipulatives—Spinners website and Random Birthday Applet website listed below.
- Explore more about the irrational number pi by visiting the Pi websites listed below.
- Share other works of Arp, including his writings and sculptures, by using the book *Arp* or by viewing the Jean Arp websites below.
- Explore the period and work of other dada artists using the Dadaism websites listed below.

Cross-Curricular Connections:

Science

- Research and discuss the important role of probability in genetics. Using a Punnet square, determine the probabilities and possible outcomes of eye color in humans.

Social Studies

- Research and discuss the causes, major events, duration, and aftermath of World War I.

Related Literature

Arp, J. (1972). *Arp on Arp.* New York: Viking Adult.

Blythe, S., & Powers, E. (2006). *Looking at dada.* New York: The Museum of Modern Art.

Fauchereau, S. (1988). *Arp.* New York: Rizzoli.

Lippard, L. (1971). *Dadas on art: Tzara, Arp, Duchamp and others.* Mineola, NY: Dover.

Museum of Modern Art. (1986). *Arp.* Stratford, NH: Ayer.

Related Instructional Resources

Aigner-Clark, J. (2004). *Baby Einstein: The ABCs of art.* New York: Hyperion Books for Children.

Dickins, R. (2005). *The children's book of art: An introduction to famous paintings.* London: Usborne.

Evans, J., & Skelton, T. (2001). *How to teach art to children.* Monterey, CA: Evan-Moor.

Gale, M. (1997). *Dada and surrealism A&I (art and ideas).* New York: Phaidon Press.

Klar, A. (2007). *Surreal people.* London: V & A.

Kohl, M., & Solga, K. (1996). *Discovering great artists: Hands-on art for children in the styles of the great masters.* Bellingham, WA: Bright Ring.

Krull, K. (1995). *Lives of the artists: Masterpieces, messes.* San Diego: Harcourt Brace.

Pappas, T. (1989). *The joy of mathematics.* San Carlos, CA: World Wide Publishing/Tetra.

Renshaw, A., & Ruggi, G. (2005). *The art book for children.* New York: Phaidon Press.

Scieszka, J., & Smith, L. (2005). *Seen art?* New York: Viking Press.

Related Websites

Buffon's Needle
http://www.angelfire.com/wa/hurben/buff.html
http://www.mste.uiuc.edu/reese/buffon/buffon.html

Dadaism
http://www.artlex.com/ArtLex/d/dada.html
http://www.artcyclopedia.com/history/dada.html

History of Art Timeline
http://www.printsforpleasure.com/historyofart.htm

Jean Arp
http://www.artchive.com/artchive/A/arp.html
http://www.artcyclopedia.com/artists/arp_jean.html
http://www.virtualology.com/virtualmuseumofart/hallofsculpture/JEANARP.COM/

Jean Arp's *Collage with Squares Arranged According to the Laws of Chance*
http://www.moma.org/collection/browse_results.php?object_id=37013

Major Modern and Contemporary Visual Artists
http://www.the-artists.org/index.cfm

Pi
http://www.facade.com/legacy/amiinpi/
http://www.factmonster.com/ce6/sci/A0838910.html
http://newton.ex.ac.uk/research/qsystems/collabs/pi/

Random Birthday Applet
http://www-stat.stanford.edu/~susan/surprise/Birthday.html

Study Art
http://www.sanford-artedventures.com/study/study.html

Teach Art
http://www.sanford-artedventures.com/teach/teach.html

Virtual Manipulatives—Coin Tossing
http://nlvm.usu.edu/en/nav/frames_asid_305_g_3_t_5.html

Virtual Manipulatives—Spinners
http://nlvm.usu.edu/en/nav/frames_asid_186_g_3_t_5.html?open=activities

Paul Klee (The Life and Work of . . . Series) (2000)

by Sean Connelly

Heinemann Library, ISBN #1403404992

Paul Klee (Getting to Know the World's Greatest Artists Series) (1991)

by Mike Venezia

Children's Press, ISBN #0516422944

Overview of Books: Enjoy both of these biographies of the twentieth-century Swiss-German abstract artist, Paul Klee, whose work was influenced by surrealist and cubist painters and whose works often featured letters, numerals, and simple shapes, making his art appealing to both children and adults.

Mathematical Concepts and Skills: four-color map theorem, problem solving, estimation, graph theory

Visual Arts Concepts and Skills: line, shape, color, value, creating art (watercolored name strip), art history (life and artwork of Paul Klee)

Overview of Activities: Students explore and learn about the life and artwork of Paul Klee and create colored name strips in the spirit of his masterpiece, *Once Emerged from the Gray of the Night*. Students then use the work of Paul Klee as a springboard into the exploration of the famed Four-Color Theorem.

National Mathematics Standards (2000): Students in grades 6–8 should "build new mathematical knowledge through problem solving," "solve problems that arise in mathematics and in other contexts," and "apply and adapt a variety of appropriate strategies to solve problems" (p. 402) (Problem Solving Standard). Students in grades 6–8 should also "recognize and apply geometric ideas and relationships in areas outside of the mathematics classroom, such as art, science, and everyday life" (p. 397) (Geometry Standard).

National Visual Arts Standards (1994): Students in grades 5–8 should choose and evaluate a range of subject matter, symbols, and ideas (Content Standard 3) and "use subjects, themes, and symbols that demonstrate knowledge of contexts, values, and aesthetics that communicate intended meaning in artworks" (p. 50) (Achievement Standard b); understand the visual arts in relation to history and cultures (Content Standard 4) and "analyze, describe, and demonstrate how factors of time and place influence visual characteristics that give meaning and value to a work of art" (p. 50) (Achievement Standard c); reflect upon and assess the characteristics and merits of their work and the work of others (Content Standard 5) and "describe and compare a variety of individual responses to their own artworks and to artworks from various eras and cultures" (p. 51) (Achievement Standard c); make connections between visual arts and other disciplines (Content Standard 6) and "describe ways in which the principles and subject matter of other disciplines taught in the school are interrelated to the visual arts" (p. 51) (Achievement Standard b).

Materials: images of Klee's works (e.g., *Fugue in Red, Crystal Gradation, Ad Parnassum, Senecio, Once Emerged from the Gray of the Night, Polyphonically Framed White*), 9" × 12" posterboard (precut prior to class into three paper strips each measuring 4" × 12"), black crayons, watercolor paints, blank map of the United States, colored pencils or crayons

Description of Activities:

1. Share with students images of Klee's artwork as well as select passages about the life of Paul Klee from *The Life and Work of Paul Klee* and *Paul Klee* (Getting to Know the World's Greatest Artists Series). Paul Klee was a twentieth-century Swiss-German abstract artist as well as a talented violinist. Many of his works portray his love of music; in fact, his use of patterns and color in his paintings (e.g., *Fugue in Red* and *Crystal Gradation*) seem to imitate the rhythm of music. Influenced by the expressionist artist Wassily Kandinsky and the cubist artist Pablo Picasso, many of Klee's works feature colored squares (e.g., *Ad Parnassum, Senecio,* and *Polyphonically Framed White*). Klee often used simple lines and shapes as well as letters and numbers in his works (e.g., *Once Emerged from the Gray of the Night*), which made his works appealing to children and adults.
2. Facilitate a discussion about what features students find appealing (or unappealing) about Klee's works. Do they see the cubist influence in his work? The influence of expressionist painters? How does his artwork differ from other periods of art such as impression, pointillism, pop art, etc.?
3. Show the artwork *Once Emerged from the Gray of the Night.* Encourage students to notice how Klee chose to shade the various regions around each letter with different colors and shades of colors.

4. Prior to class, take several pieces of 9" × 12" white posterboard and cut each piece into two strips each measuring 4" × 12" (discard the left-over 1" × 12" strip). Distribute to each student one rectangular strip. Using a black crayon, students write their first name on the strip in large block letters, filling the strip.
5. In the spirit of Klee, students use watercolors to paint the resulting white regions of each letter in various colors. Students should select their colors such that adjacent (touching) regions are a different color. Students should experiment with the value of color and focus on selecting colors that result in an aesthetically pleasing name strip.
6. Create a rectangular mural of the students' name strips by hanging them on a wall in the classroom.
7. Students observe the name strips and compare and discuss the most appealing (or unappealing) uses of color appearing in particular name strips.
8. Revisit Klee's *Once Emerged from the Gray of the Night,* reminding students how Klee chose to shade the various regions around each letter with different colors and shades of colors, as in their name strips.
9. Place students into small groups. Distribute a blank map of the United States to students. Ask students to view the map and to estimate the fewest number of different colored crayons that would be needed to color the map such that adjacent regions (i.e., regions that share a boundary) are different colors. Allow time within groups for students to share their thoughts and to justify their reasoning. After coming to a consensus, let the small groups share their predictions with the entire class and defend their reasoning.
10. Distribute colored pencils or crayons to the groups. Students work collaboratively coloring the map to determine the fewest number of colors needed to color the map as described. (Four colors are needed.)
11. After each group has determined a solution, students share their work and describe their strategies for solving the problem.

12. Visit the online book *The Story of the Young Map Colorer*, using the link listed below, which is a short tale in which readers encounter the centuries-old mathematical conundrum commonly referred to as the Four-Color Theorem that has challenged both mapmakers and mathematicians. For centuries, it was a commonly known, but unproven, rule among mapmakers that a map drawn on a flat surface or sphere requires no more than four colors to show distinct regions. Having regions of distinct colors was important, as a mapmaker would want the user to easily discern adjacent regions. The Four-Color Theorem was solved in 1976 with the aid of a computer, and researchers affirmed that no more than four colors would be needed to color any map, no matter how intricate or complex.
13. Challenge students to draw a map that needs only two colors and one that requires only three colors. Can they create a map that needs five colors?

Assessment:

- Did students articulate the features they liked and disliked in Klee's works?
- Did students create a name strip in the spirit of Klee's *Once Emerged from the Gray of the Night*?
- Did students provide reasonable solutions to the map-coloring problem and defend their reasoning?
- Did students create maps needing only two colors? Three colors?

Activity Extensions:

- Students research and report on the field of mathematics known as graph theory. Who were the major mathematicians associated with graph theory?
- Students research the history of map making.

Cross-Curricular Connections:

Science

- Students locate areas on a map that have experienced constructive forces (e.g., volcanic eruptions, deposition of sediment, crustal formation) and destructive forces (e.g., weathering and erosion).

Social Studies

- Students identify major mountain ranges, rivers, and other pertinent geographic features on a map.

Related Literature

Connelly, S. (2000). *Paul Klee* (The life and work of . . . series). Chicago: Heinemann Library.

Laidlaw, J. (2001). *Paul Klee* (Artists in their time series). New York: Scholastic.

Venezia, M. (1991). *Paul Klee* (Getting to know the world's greatest artists series). Chicago: Children's Press.

Related Instructional Resources

Aigner-Clark, J. (2004). *Baby Einstein: The ABCs of art.* New York: Hyperion Books for Children.

Dickins, R. (2005). *The children's book of art: An introduction to famous paintings.* London: Usborne.

Evans, J., & Skelton, T. (2001). *How to teach art to children.* Monterey, CA: Evan-Moor.

Kohl, M., & Solga, K. (1996). *Discovering great artists: Hands-on art for children in the styles of the great masters.* Bellingham, WA: Bright Ring.

Krull, K. (1995). *Lives of the artists: Masterpieces, messes.* San Diego: Harcourt Brace.

Renshaw, A., & Ruggi, G. (2005). *The art book for children.* New York: Phaidon Press.

Scieszka, J., & Smith, L. (2005). *Seen art?* New York: Viking Press.

Ward, R. (Spring 2006). Modeling effective pedagogical strategies for teaching mathematics. *The Charter Schools Resource Journal,* 1–9.

Related Websites

Four-Color Theorem and Map Coloring
http://www.c3.lanl.gov/mega-math/workbk/map/mpbkgd.html

History of Art Timeline
http://www.printsforpleasure.com/historyofart.htm

Major Modern and Contemporary Visual Artists
http://www.the-artists.org/index.cfm

Paul Klee
http://www.ibiblio.org/wm/paint/auth/klee/
http://www.artcyclopedia.com/artists/klee_paul.html
http://www.abcgallery.com/K/klee/klee.html

Paul Klee Interactive Puzzle
http://www.swissinfo.org/flash/special/klee_2005/eng/pk.html

Paul Klee's *Once Emerged from the Gray of the Night*
http://cgfa.sunsite.dk/klee/p-klee6.htm

The Story of the Young Map Colorer
http://www.c3.lanl.gov/mega-math/workbk/map/mpprstory.html

Study Art
http://www.sanford-artedventures.com/study/study.html

Teach Art
http://www.sanford-artedventures.com/teach/teach.html

Assessment Resources References

Bright, G., & Joyner, J. (1998). *Classroom assessment in mathematics.* Lanham, MD: University Press of America.

British Columbia Ministry of Education. (n.d.). *Handbook of assessment tools for process evaluation in mathematics.* Victoria, BC: The Ministry.

Bryant, D., & Driscoll, M. (1998). *Exploring classroom assessment in mathematics: A guide for professional development.* Reston, VA: NCTM.

Burrill, J. (Ed.). (2005). *Mathematics assessment sampler: Grades 6–8.* Reston, VA: NCTM.

Bush, W. (2000). *Mathematics assessment: Cases and discussion questions for grades 6–12.* Reston, VA: NCTM.

Bush, W., & Leinwand, S. (2001). *Mathematics assessment: A practical handbook for grades 6–8.* Reston, VA: NCTM.

Charles, R., Lester, F., & O'Daffer, P. (1987). *How to evaluate progress in problem solving.* Reston, VA: NCTM.

Clarke, D. (1997). *Constructive assessment in mathematics: Practical steps for classroom teachers.* Emeryville, CA: Key Curriculum Press.

Kulm, G. (1994). *Mathematics assessment: What works in the classroom.* San Francisco: Jossey-Bass.

Madfes, T. (1999). *Learning from assessment: Tools for examining assessment through standards.* Reston, VA: NCTM.

Mathematical Sciences Education Board. (1994). *Measuring up: Prototypes for mathematics assessment.* Washington, DC: National Academy Press.

National Council of Teachers of Mathematics (NCTM). (1995). *Assessment standards for school mathematics.* Reston, VA: NCTM

Ott, J. (1994). *Alternate assessment in the mathematics classroom.* Columbus, OH: Glencoe/McGraw-Hill.

Pandey, T. (1991). *A sampler of mathematics assessment.* Sacramento: California Department of Education.

Schoenfeld, A. (1999). *Middle grades assessment.* White Plains, NY: Dale Seymour.

Stenmark, J. (1989). *Assessment alternatives in mathematics: An overview of assessment techniques that promote learning.* Berkeley: University of California.

Stenmark, J. (Ed.) (1991). *Mathematics assessment: Myths, models, good questions, and practical suggestions.* Reston, VA: NCTM.

Stenmark, J., & Bush, W. (2001). *Mathematics assessment: A practical handbook for grades 3–5.* Reston, VA: NCTM.

Walch, J. (2003). *Assessment: Strategies for math.* Portland, ME: Walch.

Webb, N., & Coxford, A. (1993). *Assessment in the mathematics classroom.* Reston, VA: NCTM.

Literature References

Adler, D. (1991). *A picture book of Christopher Columbus.* New York: Scholastic.

Alloway, L. (1983). *Roy Lichtenstein.* New York: Abbeville Press.

Anderson, M. (1995). *Amazing Leonardo da Vinci inventions you can build yourself.* Chicago: Nomad Books.

Anno, M. (1999). *Anno's mysterious multiplying jar.* New York: Putnam Books.

Antoine, V. (1994). *Picasso: A day in his studio* (Art for children series). New York: Chelsea House.

Appel, J., & Guglielmo, A. (2006). *Pop Warhol's top.* New York: Sterling.

Arnold, T. (2000). *Parts.* New York: Puffin Books.

Arnold, T. (2005). *More parts.* New York: Puffin Books.

Arnold, T. (2007). *Even more parts.* New York: Puffin Books.

Arp, J. (1972). *Arp on Arp.* New York: Viking Adult.

Barasch, L. (2005). *Ask Albert Einstein.* New York: Frances Foster Books.

Barber, J. (2000). *President* (Eyewitness Books). New York: Dorling Kindersley.

Bargallo, E. (2006). *My name is Picasso.* Hauppauge, NY: Barron's Educational Series.

Barretta, G. (2006). *Now and Ben: The modern inventions of Ben Franklin?* New York: Holt.

Bender, L. (2005). *Invention* (Eyewitness books series). New York: Dorling Kindersley.

Bentley, D. (2002). *The seven wonders of the ancient world.* London: Oxford University Press.

Berg, A., &, Bochner, A. (2002). *The totally awesome business book for kids.* New York: Newmarket Press.

Berg, A., &, Bochner, A. (2002). *The totally awesome money book for kids.* New York: Newmarket Press.

Bingham, C. (2004). *Invention* (Eye Wonder Series). New York: Dorling Kindersley.

Bingham, C. (2004). *Pyramid* (DK Eye Wonder Series). New York: Dorling Kindersley.

Birch, D. (1988). *The king's chessboard.* New York: Puffin Books.

Blanquet, C. (1993). *Miro: Earth and sky* (Art for children series). New York: Chelsea House.

Blythe, S., & Powers, E. (2006). *Looking at dada.* New York: The Museum of Modern Art.

Bramwell, M. (1998). *How maps are made* (Maps and mapmakers series). Minneapolis: Lerner.

Bramwell, M. (1998). *Mapping in everyday life* (Maps and mapmakers series). Minneapolis: Lerner.

Bramwell, M. (1998). *Mapping our world* (Maps and mapmakers series). Minneapolis: Lerner.

Branley, F. (1973). *Eclipse: Darkness in daytime.* New York: Crowell.

Branley, F. (1976). *How little and how much.* New York: Crowell.

Branley, F. (1998). *The planets in our solar system.* New York: HarperCollins.

Brewer, D. (2004). *Inventions* (100 things you should know about series). New York: Barnes & Noble Books.

Britton, T. (2002). *The World Trade Center* (Symbols, landmarks and monuments series). Edina, MN: ABDO.

Brooks, P. (1995). *Georgia O'Keeffe: An adventurous spirit.* Chicago: Franklin Watts.

Brown, D. (1998). *Odd boy out: Young Albert Einstein.* Boston: Houghton Mifflin.

Brumbeau, J. (2001). *The quiltmaker's gift.* New York: Scholastic.

Brumbeau, J. (2004). *The quiltmaker's journey.* New York: Orchard Books.

Bryant, J. (2005). *Georgia's bones.* Grand Rapids, MI: Eerdmans Books for Young Readers.

Burleigh, R. (2006). *Paul Cezanne: A painter's journey.* New York: Abrams.

Campbell, J., & Collison, C. (2005). *G is for galaxy: An out of this world alphabet*. Chelsea, MI: Sleeping Bear Press.

Carlisle, M. (1991). *Let's investigate marvelously meaningful maps*. New York: Barron's Educational Series.

Challoner, J. (1997). *Hot and cold*. Austin, TX: Steck-Vaughn.

Cheney, L. (2005). *A time for freedom: What happened when in America*. New York: Simon & Schuster.

Chrisp, P. (2006). *Pyramid* (DK experience series). New York: Dorling Kindersley.

Clements, A. (2006). *A million dots*. New York: Simon & Schuster Books for Young Readers.

Coerr, E. (1997). *Sadako*. New York: Penguin Young Readers.

Coerr, E. (2004). *Sadako and the thousand paper cranes*. New York: Penguin Young Readers.

Connelly, S. (2000). *Paul Klee* (The life and work of . . . series). Chicago: Heinemann Library.

Connelly, S. (2006). *Paul Cezanne* (The life and work of . . . series). Chicago: Heinemann Library.

Corrbishley, M. (1996). *The world of architectural wonders*. New York: Bedrick Books.

Crampton, W. (2002). *Smithsonian handbooks: Flags*. New York: Dorling Kindersley.

Crampton, W. (2003). *Spotter's guide to flags of the world*. London: Usborne.

Cribb, J. (2005). *Money* (Eyewitness books series). New York: Dorling Kindersley.

Davis, G. (2004). *Wackiest White House pets*. New York: Scholastic.

Davis, K. (2002). *Don't know much about the presidents*. New York: HarperCollins.

DeMauro, L. (2002). *Thomas Edison: A brilliant inventor* (Time for kids series). New York: HarperCollins.

Demi. (1997). *One grain of rice*. New York: Scholastic.

Egan, L. (1999). *Inventors and inventions*. New York: Scholastic.

Ellis, J. (2004). *What's your angle, Pythagoras?* Watertown, MA: Charlesbridge.

Escher, M. (2004). *M. C. Escher: The graphic work*. Hohenzollernring, Germany: Taschen.

Faerna, J. (1997). *Mondrian*. New York: Adams.

Fauchereau, S. (1988). *Arp*. New York: Rizzoli.

Fifth-Grade Students of Holly Springs Elementary School, Pickens, SC. (2000). *Perfect inventions = perfect world*. New York: Scholastic.

Firefly Books. (2003). *Firefly guide to flags of the world*. Richmond Hills, Ontario: Firefly Books.

Fisher, L. (2005). *Marie Curie*. New York: Macmillan.

Freedman, R. (1981). *Animal superstars: Biggest, strongest, fastest, smartest*. Upper Saddle River, NJ: Prentice-Hall.

Friedman, A. (1994). *A cloak for the dreamer*. New York: Scholastic.

Froman, R. (1974). *A game of functions*. New York: Crowell.

Fuqua, N. (2004). *First pets: Presidential best friends*. New York: Scholastic.

Garrison, J., & Tubesing, A. (1996). *A million visions of peace: Wisdom from the Friends of Old Turtle*. New York: Scholastic.

Gerstein, M. (2003). *The man who walked between the towers*. Brookfield, CT: Roaring Brook Press.

Gibbons, G. (2004). *Mummies, pyramids, and pharaohs: A book about ancient Egypt*. New York: Little, Brown.

Gibbons, G. (2005). *Planets*. New York: Holiday House.

Goldsmith, M. (2001). *Galileo Galilei* (Scientists who made history series). Austin, TX: Raintree Steck-Vaughn.

Graham, I. (2006). *Curie and the science of radioactivity*. New York: Barron's Educational Series.

Graham-Barber, L. (1992). *Doodle dandy! The complete book of independence day words*. New York: Bradbury Press.

Grange Books. (2005). *Mondrian*. London: Grange Books.

Greenberg, J., & Jordan, S. (1994). *The American eye: Eleven artists of the twentieth century.* New York: Delacorte Press.

Haber, L. (1992). *Black pioneers of science and invention.* San Diego: Harcourt Children's Books.

Harman, H. (2004). *Money sense for kids*. Hauppauge, NY: Barron's Educational Series.

Harper, C. (2001). *Imaginative inventions: The who, what, when, where, and why of roller skates, potato chips, marbles and pie (and more!)*. Boston: Little, Brown.

Harris, N. (2003). *Paul Cezanne* (Artists in their time series). New York: Scholastic.

Hickey, D., & Brown, E. (2005). *Roy Lichtenstein: Prints from the Jordan D. Schnitzer Collection.* Seattle: Marquand Books.

Hoose, P. (2002). *It's our world, too!* New York: Farrar Straus Giroux.

Howe, R. (2002). *Flags of the fifty states and their incredible histories: The complete guide to America's most powerful symbols.* Guilford, CT: Lyons Press.

Hyman, T. (2006). *Pyramids* (Wonders of the world series). Farmington Hills, MI: Kidhaven Press.

Ishii, T. (2001). *One thousand paper cranes: The story of Sadako and the children's peace statue*. New York: Random House Children's Books.

Jenkins, S. (1995). *Biggest, strongest, fastest*. Boston: Houghton Mifflin.

Jenkins, S. (2004). *Actual size*. Boston: Houghton Mifflin.

Jenkins, S. (2005). *Prehistoric actual size*. Boston: Houghton Mifflin.

Jenkins, S. (2006). *Almost gone: The world's rarest animals*. New York: Scholastic.

Jenkins, S., & Page, R. (2003). *What do you do with a tail like this?* Boston: Houghton Mifflin.

Jonas, A. (1987). *Reflections*. New York: Greenwillow Books.

Jones, L. (2000). *Five brilliant scientists* (Great black heroes series). New York: Scholastic.

Kelley, T. (2002). *Pablo Picasso: Breaking all the rules* (Smart about art series). New York: Grosset & Dunlap.

Kerrod, R. (2003). *Universe* (Eyewitness books series). New York: Dorling Kindersley.

Kincade, S. (1992). *Our time is now* (Young people changing the world). Upper Saddle River, NJ: Pearson Foundation.

Knowlton, J. (1985). *Maps and globes*. New York: HarperCollins.

Knowlton, J. (1988). *Geography from A to Z: A picture glossary*. New York: HarperCollins.

Kramer, S. (2004). *Who was Ferdinand Magellan?* (Who was . . . ? series). New York: Penguin Young Readers.

Krull, K. (1995). *Leonardo da Vinci.* New York: Penguin Books.

LaFontaine, B. (1998). *Great inventors and inventions* (100 things you should know about series). London: Dover.

Laidlaw, J. (2001). *Paul Klee* (Artists in their time series). New York: Scholastic.

Laliberte, M. (2005). *Seven wonders of the ancient world.* Berkeley Heights, NJ: Enslow.

Lasky, K. (1994). *The librarian who measured the earth.* Boston: Little, Brown.

Lassieur, A. (1998). *Marie Curie: A scientific pioneer* (Great life stories series). New York: Franklin Watts.

Lemaire, G. (1997). *Calder* (Great modern masters series). New York: Abrams.

Levy, D. (2003). *Stars and planets* (Discoveries series). New York: Barnes & Noble Books.

Levy, D. (2005). *The World Trade Center* (Great structures in history series). New York: Thomson Gale.

Lewis, B. (1992). *Kids with courage: True stories about young people making a difference.* Minneapolis: Free Spirit.

Lewis, J. (1998). *Doodle dandies: Poems that take shape*. New York: Aladdin.

Lewis, J. (2002). *A world of wonders: Geographic travels in verse and rhyme.* New York: Dial Books for Young Readers.

Lewis, J. (2005). *Galileo's universe*. Monroe, WI: The Creative Company.

Lippard, L. (1971). *Dadas on art: Tzara, Arp, Duchamp and others.* Mineola, NY: Dover.

Lippincott, K. (2004). *Astronomy* (Eyewitness books series). New York: Dorling Kindersley.

Ljungkvist, L. (2006). *Follow the line*. New York: Penguin Group.

Lowery, L. (1996). *Georgia O'Keeffe*. Minneapolis: Carolrhoda Books.

Lowery, L. (2006). *Pablo Picasso.* Minneapolis: Lerner.

Lynch, A. (2005). *Great buildings* (Discoveries series). New York: Barnes & Noble Books.

Macaulay, D. (1975). *Pyramid.* Boston: Houghton Mifflin.

Macaulay, D. (1981). *Cathedral: The story of its construction.* Boston: Houghton Mifflin.

Macaulay, D. (1982). *Castle.* Boston: Houghton Mifflin.

Macaulay, D. (1983). *City: A story of Roman planning and construction.* Boston: Houghton Mifflin.

Macaulay, D. (2004). *Building big.* Boston: Houghton Mifflin.

Maestro, B. (1999). *The story of clocks and calendars.* New York: HarperCollins.

Malet, R. (2003). *Joan Miro.* New York: Rizzoli.

Maltbie, P. (2005). *Picasso and Minou.* Watertown, MA: Charlesbridge.

Marter, J. (1997). *Alexander Calder.* Cambridge: Cambridge University Press.

Martin, J. (1998). *Snowflake Bentley*. Boston: Houghton Mifflin.

Maruki, T. (1982). *Hiroshima no pika*. New York: HarperCollins.

Marzollo, J. (1994). *My first book of biographies: Great men and women every child should know.* New York: Scholastic.

Mason, A. (1994). *Leonardo da Vinci: An introduction to the artist's life and work* (Barron's famous artist series). London: Aladdin Books.

Mason, A. (1994). *Pablo Picasso* (Famous artists series). London: Aladdin Books.

Mason, A. (1994). *Paul Cezanne* (Famous artists series). Hauppauge, NY: Barron's Educational Series.

Mason, A. (2004). *Leonardo da Vinci* (Lives of the artists series). Milwaukee: Gareth Stevens.

Matthews, R. (2005). *Explorers* (Eyewitness books series). New York: Dorling Kindersley.

Mayr, D. (2002). *Kids' everything money: From saving to spending to investing—Learn all about money!* Avon, MA: Adams Media.

McCallum, A. (2005). *Beanstalk: The measure of a giant.* Watertown, MA: Charlesbridge.

McCullough, N. (2006). *Essential book of presidential trivia.* New York: Random House.

Meltzer, M. (1993). *Gold: The true story of why people search for it, mine it, trade it, fight for it, mint it, display it, steal it, and kill for it.* New York: HarperCollins.

Merriam Webster. (2003). *Webster's international atlas.* New York: Barnes & Noble.

Miller, D. (2003). *Arctic nights arctic lights.* New York: Walker.

Milner, J. (1995). *Mondrian.* New York: Phaidon Press.

Mines, C. (2006). *Cubism.* Farmington Hills, MI: Thomas Gale.

Mink, J. (2000). *Miro.* Hohenzollernring, Germany: Taschen.

Morley, J., Bergin, M., & James, J. (1991). *An Egyptian pyramid.* New York: Bedrick Books.

Museum of Modern Art. (1986). *Arp*. Stratford, NH: Ayer.

Parker, S. (1995). *Marie Curie and radium* (Science discoveries series). New York: Chelsea House.

Pershing Accelerated School Students. (2002). *We dream of a world.* New York: Scholastic.

Pettenati, J, J. (2006). *Galileo's journal, 1609–1610.* Watertown, MA: Charlesbridge.

Prelutsky, J. (1996). *A pizza the size of the sun.* New York: Greenwillow Books.

Putnam, J. (2004). *Mummy* (Eyewitness books series). New York: Dorling Kindersley.

Putnam, J., Hayman, P., & Brightling, G. (2004). *Pyramid* (Eyewitness books series). New York: Dorling Kindersley.

Reichblum, C. (2004). *Dr. Knowledge presents: Strange & fascinating facts about the presidents.* New York: Black Dog & Leventhal.

Reimer, L., & Reimer, W. (1995). *Mathematicians are people, too.* Palo Alto, CA: Dale Seymour.

Roca, N. (2002). *Boys and girls of the world: From one end to the other.* Hauppauge, NY: Barron's Educational Series.

Ross, N. (1995). *Miro* (Famous artists series). Hauppauge, NY: Barron's Educational Series.

Roza, G. (2005). *An optical artist: Exploring patterns and symmetry.* New York: PowerKids Press.

Rubel, D. (1994). *Scholastic encyclopedia of the presidents and their times.* New York: Scholastic.

Ryan, P. (1989). *Explorers and mapmakers.* New York: Lodestar Books.

Satterfield, K. (2005). *Benjamin Franklin: A man of many talents* (Time for kids series). New York: HarperCollins.

Scarborough, K. (2001). *Pablo Picasso* (Artists in their times series). New York: Scholastic.

Schaefer, A. (2003). *Alexander Calder* (The life and works of . . . series). Chicago: Heinemann Library.

Schonberg, M. (2005). *I is for idea: An inventions alphabet.* Chelsea, MI: Sleeping Bear Press.

Schwartz, D. (2003). *Millions to measure.* New York: HarperCollins.

Scieszka, J., & Smith, L. (1995). *Math curse.* New York: Penguin Books.

Scieszka, J., & Smith, L. (2004). *Science verse.* New York: Penguin Books.

Scillian, D. (2003). *P is for passport.* Chelsea, MI: Sleeping Bear Press.

Sellier, M. (1996). *Cezanne from A to Z.* New York: Bedrick Books.

Shields, C. (1998). *Lunch money and other poems about school.* New York: Puffin Books.

Silverstein, S. (1981). *A light in the attic.* New York: HarperCollins.

Silverstein, S. (1996). *Falling up.* New York: HarperCollins.

Silverstein, S. (2004). *Where the sidewalk ends.* New York: HarperCollins.

Sipiera, P. (1991). *Globes* (A new true book series). Chicago: Children's Press.

Sis, P. (1996). *Starry messenger.* New York: Farrar Straus Giroux.

Smith, D. (2002). *If the world were a village: A book about the world's people.* Tonawanda, NY: Kids Can Press.

Smith, D. (2003). *Mapping the world by heart.* Watertown, MA: Snyder.

Smith, W. (2001). *Flag lore of all nations.* Brookfield, CT: Millbrook Press.

Sobel, S. (1999). *How the U.S. government works.* Hauppauge, NY: Barron's Educational Series.

Sobel, S. (2001). *Presidential elections: And other cool facts.* Hauppauge, NY: Barron's Educational Series.

Spier, P. (1980). *People.* New York: Doubleday.

Steele, P. (2005). *Galileo: The genius who faced the inquisition.* Washington, DC: National Geographic Society.

Steinke, A. (1987). *Marie Curie and the discovery of radium* (Solutions series). New York: Barron's Educational Series.

Stephens, P. (2001). *Tessellations: The history and making of symmetrical design.* Aspen, CO: Crystal.

St. George, J. (2002). *So you want to be an inventor?* New York: Penguin.

Sullivan, G. (1987). *Facts and fun about the presidents.* New York: Scholastic.

Taylor, B. (1993). *Maps and mapping* (Young discoveries series). New York: Kingfisher.

Thimmesh, C. (2000). *Girls think of everything: Stories of ingenious inventions by women.* Boston: Houghton Mifflin.

Thimmesh, C. (2002). *The sky's the limit: Stories of discovery by women and girls.* Boston: Houghton Mifflin.

Thimmesh, C. (2004). *Madame President: The extraordinary, true, (and evolving) story of women in politics.* Boston: Houghton Mifflin.

Tracy, K. (2004). *Pierre and Marie Curie and the discovery of radium* (Uncharted, unexplored and unexplained scientific advancements of the 19th century series). Hockessin, DE: Mitchell Lane.

Tsuchiya, Y. (1997). *Faithful elephants: A true story of animals, people and war.* Boston: Houghton Mifflin.

Turvey, P. (1994). *Inventions: Inventors and ingenious ideas* (Timelines series). London: Franklin Watts.

VanCleave, J. (2004). *Scientists through the ages.* Hoboken, NJ: Wiley & Sons.

Venezia, M. (1988). *Paul Cezanne* (Getting to know the world's greatest artists series). Chicago: Children's Press.

Venezia, M. (1988). *Picasso* (Getting to know the world's greatest artists series). New York: Scholastic.

Venezia, M. (1991). *Paul Klee* (Getting to know the world's greatest artists series). Chicago: Children's Press.

Venezia, M. (1993). *Georgia O'Keeffe* (Getting to know the world's greatest artists series). Chicago: Children's Press.

Venezia, M. (2002). *Georges Seurat* (Getting to know the world's greatest artists series). New York: Scholastic.

Venezia, M. (2002). *Roy Lichtenstein* (Getting to know the world's greatest artists series). Chicago: Children's Press.

Walker, J. (2003). *100 things you should know about ancient Egypt.* Broomall, PA: Mason Crest.

Walker, L. (1994). *Roy Lichtenstein: The artist at work.* New York: Lodestar Books.

Walpole, B. (1995). *Temperature.* Chicago: Gareth Stevens.

Wells, R. (2003). *How do you know what time it is?* Morton Grove, IL: Whitman.

Westray, K. (2003). *Scholastic atlas of the world.* New York: Scholastic.

Whelan, G. (2001). *Homeless bird.* New York: HarperCollins.

Wood, R. (2003). *Great inventions* (Discoveries series). New York: Barnes & Noble Books.

Worth, R. (2003). *The great empire of China and Marco Polo in world history.* Berkeley Heights, NJ: Enslow.

Wulffson, D. (2001). *Kid who invented the trampoline: More surprising stories about inventions.* New York: Penguin Young Readers Group.

Znamierowski, A. (2001). *The world encyclopedia of flags.* London: Lorenz Books.

Instructional Resources References

Recommended Book Series

100 Things You Should Know About Series (Barnes & Noble Books)

Artists in Their Times Series (Scholastic)

Barron's Famous Artist Series (Aladdin Books)

Childhood of Famous Americans Series (Aladdin)

Discover America State by State Alphabet Series (Sleeping Bear Press)

Discoveries Series (Barnes & Noble Books)

DK Experience Series (Dorling Kindersley)

Don't Know Much About Series (HarperTrophy)

Eyewitness Books Series (Dorling Kindersley)

Eyewitness Visual Dictionaries Series (Dorling Kindersley)

Eye Wonder Books Series (Dorling Kindersley)

Getting to Know the World's Greatest Artists Series (Children's Press)

Giants of Science Series (Penguin Young Readers)

History Maker Bio Series (Lerner)

How Government Works Series (Lerner)

Inventions That Shaped the World Series (Scholastic)

Inventor and Inventions Series (Benchmark Books)

Once upon America Series (Puffin Books)

Scientists Who Made History Series (Raintree Steck-Vaughn)

Smart About Series (Grosset & Dunlap)

Smart About Art Series (Grosset & Dunlap)

Spend the Day In Series (Jossey-Bass)

Time for Kids Series (HarperCollins)

Timelines Series (Franklin Watts)

True Books: American Indian Series (Children's Press)

Who Was . . . ? Series (Penguin Young Readers)

Aigner-Clark, J. (2004). *Baby Einstein: The ABCs of art*. New York: Hyperion Books for Children.

Anderson, M. (2006). *Amazing Leonardo da Vinci inventions you can build yourself.* White River Junction, VT: Nomad Press.

Bando, I. (1995). *Geometry and fractions with tangrams.* Vernon Hills, IL: Learning Resources.

Bateman, K. (2001). *The young investor: Projects and activities for making your money grow.* Chicago: Chicago Review Press.

Becklake, S. (1993). *The visual dictionary of the universe* (Eyewitness visual dictionaries series). New York: Dorling Kindersley.

Bender, L. (2005). *Invention* (Eyewitness books series). New York: Dorling Kindersley.

Biesty, S. (2005). *Egypt in spectacular cross-section.* New York: Scholastic.

Bingham, C. (2004). *Invention* (Eyewonder books series). New York: Dorling Kindersley.

Boursin, D. (2001). *Origami paper animals.* New York: Firefly Books.

Boutan, M. (1995). *Cezanne.* San Francisco: Raincoast Books.

Bowen, D. (1991). *The visual dictionary of the human body* (Eyewitness visual dictionaries series). New York: Dorling Kindersley.

Brewer, D. (2004). *Inventions* (100 things you should know about series). New York: Barnes & Noble Books.

Burnie, D. (2000). *Light* (Eyewitness books series). New York: Dorling Kindersley.

Burnie, D. (2004). *Bird* (Eyewitness books series). New York: Dorling Kindersley.

Casey, S. (2005). *Kids inventing! A handbook for young inventors.* San Francisco: Jossey-Bass.

Casson, A (2006). *Introduction to the human body.* Philadelphia: Chelsea House.

Challoner, J. (2000). *Energy* (Eyewitness books series). New York: Dorling Kindersley.

Chatterjee, M. (2002). *India* (Eyewitness books series). New York: Dorling Kindersley.

Chipman, D. (1998). *Cool women.* Chicago: Girl Press.

Clark, B., & Buller, L. (2005). *Amphibian* (Eyewitness books series). New York: Dorling Kindersley.

Cooper, C. (1999). *Matter* (Eyewitness books series). New York: Dorling Kindersley.

Cros, P. (2006). *Cezanne.* New York: Vilo.

Currie, S. (2001). *Women inventors.* San Diego: Lucent Books.

Dale Seymour Publications. (1991). *Tessellation winners: Escher-like original student art, the first contest.* Palo Alto, CA: Dale Seymour.

Dale Seymour. (1997). *Tessellation teaching masters.* Palo Alto, CA: Dale Seymour.

D'Alusio, F. (1998). *Women in the material world.* San Francisco: Sierra Club Books.

Dean, C. (1993). *Cezanne.* New York: Phaidon Press.

Dickins, R. (2005). *The children's book of art: An introduction to famous paintings.* London: Usborne.

Dickinson, T. (1995*). Other worlds: A beginner's guide to planets and moons.* Tonawanda, NY: Firefly.

DK Publishing. (1998). *DK nature encyclopedia.* London: DK Children.

Egan, L. (1999). *Inventors and inventions.* New York: Scholastic.

Elfers, J., & Schuyt, M. (2000). *Tangrams: 1600 ancient Chinese puzzles.* New York: Barnes & Noble.

Ellyard, D. (1996). *Weather* (Eyewitness discoveries series). New York: Dorling Kindersley.

Escher, M. (2004). *M. C. Escher: The graphic work.* Hohenzollernring, Germany: Taschen.

Evans, J., & Skelton, T. (2001). *How to teach art to children.* Monterey, CA: Evan-Moor.

Fauchereau, S. (1994). *Mondrian.* New York: Rizzoli.

Flack, J. (1989). *Inventing, inventions, and inventors: A teaching resource book.* Englewood, CO: Teacher Ideas Press.

Ford, B. E. (1990). *The master revealed: A journey with tangrams.* Vallejo, CA: Tandora's Box Press.

Ford, B. E. (1990). *Tangrams: The magnificent seven piece puzzle.* Vallejo, CA: Tandora's Box Press.

Fredericks, A., & Asimov, I. (2001). *Science fair handbook: The complete guide for teachers and parents.* Tucson: Good Year Books.

Furniss, T. (1999*). Atlas of space exploration.* Milwaukee: Gareth Stevens.

Gale, M. (1997). *Dada and surrealism A&I (art and ideas).* New York: Phaidon Press.

Ganeri, A. (1995). *Maps and mapmaking.* New York: Scholastic.

Ganeri, A. (2005). *India.* New York: Sea to Sea.

Gardner, R. (1999). *Science projects about math.* Springfield, NJ: Enslow.

Gardner, R. (2000). *Science projects about methods of measuring.* Berkeley Heights, NJ: Enslow.

Gardner, R. (2006). *Melting, freezing, and boiling science projects with matter.* Berkeley Heights NJ: Enslow.

Goodnow, J. (1994). *Math discoveries with tangrams.* Grand Rapids, MI: Ideal School Supply.

Haban, R. (1992). *How proudly they wave.* Minneapolis: Lerner.

Hahn, E. (1996). *16 extraordinary American women.* Portland, ME: Walch.

Hare, T. (2005). *Animal fact file: Head-to-tail profiles of more than 90 mammals.* New York: Checkmark Books.

Harper, C. (2001). *Imaginative inventions: The who, what, when, where, and why of roller skates, potato chips, marbles and pie (and more!).* Boston: Little, Brown.

Harrington, P. (1997). *Eclipse! The what, where, when, why & how guide to watching solar and lunar eclipses.* New York: Wiley & Sons.

Hart, G. (2003). *Ancient Egypt* (Discoveries series). New York: Barnes & Noble Books.

Hart, G. (2004). *Ancient Egypt* (Eyewitness books series). New York: Dorling Kindersley.

Hart, A., & Mantell, P. (1997). *Pyramids: 50 Hands-on activities to experience ancient Egypt.* Charlotte, VT: Williamson.

Hart, A., & Mantell, P. (2004). *Ancient Greece! 40 hands-on activities to experience this wondrous age* (Kaleidoscope kids books series). Carmel, NY: Ideals.

Holland, S. (2001). *Space* (Eye wonder series). New York: Dorling Kindersley.

Honan, L. (1999). *Spend the day in ancient Egypt: Projects and activities that bring the past to life* (Spend the day in series). New York: Jossey-Bass.

Jordan, S. (2000). *Ancient Egypt: Moments in history.* Logan, IA: Perfection Learning.

Kassinger, R. (2002). *Build a better mousetrap: Make classic inventions, discover your problem-solving genius, and take the inventor's challenge.* Hoboken, NJ: Wiley & Sons.

Kindersley, A., & Kindersley, B. (1995). *Children just like me*. New York: Dorling Kindersley.

King, D. (2007). *Have fun with the presidents: Activities, projects and fascinating facts*. Hoboken, NJ: Wiley & Sons.

Klar, A. (2007). *Surreal people*. London: V & A.

Kodansha International. (2002). *The first book of origami.* New York: Kodan-Sha America.

Kohl, M., & Solga, K. (1996). *Discovering great artists: Hands-on art for children in the styles of the great masters*. Bellingham, WA: Bright Ring.

Krull, K. (1995). *Lives of the artists: Masterpieces, messes.* San Diego: Harcourt Brace.

LaFontaine, B. (1998). *Great inventors and inventions* (100 things you should know about series). London: Dover.

Levine, S., & Johnstone, L. (2005). *First science experiments: Nature, senses, weather, & machines.* New York: Sterling.

Lewis, B. (1995). *The kid's guide to service projects: Over 500 service ideas for young people who want to make a difference.* Minneapolis: Free Spirit.

Lewis, B. (1998). *The kid's guide to social action: How to solve social problems you choose—and turn creative thinking into positive action.* Minneapolis: Free Spirit.

Lowery, L. (1996). *Georgia O'Keeffe*. Minneapolis: Carolrhoda Books.

Mackenzie, F. (1995). *Weather and seasons*. New York: Sterling.

Mason, A. (1994). *Leonardo da Vinci: An introduction to the artist's life and work* (Barron's famous artist series). London: Aladdin Books.

Mason, A. (2004). *Leonardo da Vinci* (Lives of the artists series). Milwaukee: Gareth Stevens.

Menzel, P. (1995). *Material world: A global family portrait*. San Francisco: Sierra Club Books.

Micklethwait, L. (1993). *A child's book of art: Great pictures: First words*. New York: Dorling Kindersley.

Micklethwait, L. (2004). *I spy shapes in art.* New York: Greenwillow Books.

Mogard, S. (1992). *Windows to tangrams: Reproducible activities.* Covina, CA: American Teaching Aids.

Montroll, J. (1992). *Easy origami.* Mineola, NY: Dover.

Montroll, J. (1992). *Fun with easy origami: 32 projects and 24 sheets of origami paper.* Mineola, NY: Dover.

Mound, L. (2004). *Insects* (Eyewitness books series). New York: Dorling Kindersley.

Muller, D., & Ward, R. (2007). Art and algebra? Middle school students discover algebra in Calder mobiles. *Mathematics in School, 36*(3), 15–21

National Geographic Society. (2000). *National Geographic animal encyclopedia.* Hanover, PA: National Geographic Children's Books.

National Geographic Society. (2001). *Student atlas of the world.* Washington, DC: National Geographic Society.

O'Keeffe, G. (1989). *Georgia O'Keeffe: One hundred flowers.* New York: Knopf.

Panchyk, R. (2005). *Galileo for kids: His life and ideas: 25 activities*. Chicago: Chicago Review Press.

Pappas, T. (1989). *The joy of mathematics.* San Carlos, CA: World Wide Publishing/Tetra.

Parker, S. (2000). *Medicine* (Eyewitness books series). New York: Dorling Kindersley.

Parker, S. (2004). *Mammal* (Eyewitness books series). New York: Dorling Kindersley.

Parker, S. (2005). *Fish* (Eyewitness books series). New York: Dorling Kindersley.

Pearson, A. (2004). *Ancient Greece* (Eyewitness books series). New York: Dorling Kindersley.

Phillips, C., & Priwer, S. (2006). *The everything da Vinci book: Explore the life and times of the ultimate Renaissance man*. Avon, MA: Adams.

Platt, R. (2003). *Hieroglyphics: The secrets of ancient Egyptian writing to unlock and discover* (Treasure chests series). Philadelphia: Running Press Kids.

Press, J. (2001). *Around the world art & activities: Visiting the 7 continents through craft fun*. Charlotte, VT: Williamson.

Renshaw, A., & Ruggi, G. (2005). *The art book for children*. New York: Phaidon Press.

Rose, M. (2003). *Human body* (Discoveries series). New York: Barnes & Noble.

Salvadori, M. (2000). *The art of construction: Projects and principles for beginning engineers and architects.* Chicago: Chicago Review Press.

Sammis, F. (2000). *Maps and mapmaking* (Mapping our world series). Tarrytown, NY: Marshall Cavendish.

Sarasas, C. (2002). *ABCs of origami.* North Clarendon, VT: Tuttle.

Saunders, H. (1988). *When are we ever gonna have to use this?* Palo Alto, CA: Dale Seymour.

Schoefield, L. (2003). *Ancient Greece* (Discoveries series). New York: Dorling Kindersley.

Scholastic Books. (1995). *Architecture and construction: Building pyramids, log cabins, castles, igloos, bridges, and skyscrapers.* New York: Scholastic.

Schwartz, D. (2001). *Q is for quark: A science alphabet.* Berkeley, CA: Ten Speed Press.

Scieszka, J., & Smith, L. (2005). *Seen art?* New York: Viking Press.

Shattschneider, D. (2005). *M. C. Escher: Visions of symmetry.* Petaluma, CA: Pomegranate.

Shattschneider, D., & Walker, W. (2005). *M. C. Escher kaleidocycles.* New York: Abrams.

Simon, S. (1992). *Our solar system, vol. 1.* New York: HarperCollins.

Simon, S. (1995). *The solar system: Facts and exploration*. New York: Holt.

Smith, R. (1999). *DK art school: An introduction to perspective*. New York: DK Adult.

Sobel, D. (1995). *Longitude: The true story of a lone genius who solved the greatest scientific problem of his time*. New York: Walker.

Staff of Barron's. (2004). *Our solar system: A first introduction to space and the planets.* Hauppauge, NY: Barron's Educational Series.

Stott, C., & Twist, C. (2002). *1,001 facts about space.* New York: Dorling Kindersley.

Turvey, P. (1994). *Inventions: Inventors and ingenious ideas* (Timelines series). London: Franklin Watts.

Twist, C. (2005). *Reptiles and amphibians dictionary: An A to Z of cold-blooded creatures.* New York: Scholastic.

VanCleave, J. (2004). *Scientists through the ages.* Hoboken, NJ: Wiley & Sons.

VanCleave, J. (2006). *Energy for every kid.* Hoboken, NJ: Wiley & Sons.

Van Rose, S. (2005). *Earth* (Eyewitness books series). New York: Dorling Kindersley.

Venezia, M. (1989) *Da Vinci* (Getting to know the world's greatest artists series). New York: Scholastic.

Walker, R. (2001). *Human body* (DK guide series). New York: Dorling Kindersley.

Ward, R. (2004). K–8 preservice teachers' journey into the global village: Exploring real-world data using children's literature and technology. *Arizona Reading Journal, 31*(1), 43–47.

Ward, R. (2006, Spring). Modeling effective pedagogical strategies for teaching mathematics. *The Charter Schools Resource Journal,* 1–9.

Ward, R., & Muller, D. (2006, September). Algebra and art. *Mathematics Teaching, 198,* 22–26.

Welden, A. (1998). *Girls who rocked the world: Heroines form Sacagawea to Sheryl Swoopes.* Hillsboro, OR: Beyond Words.

Williams, D. (1995). *Teaching mathematics through children's art.* Portsmouth, NH: Heinemann.

Wood, R. (2003). *Great inventions* (Discoveries series). New York: Barnes & Noble Books.

Yamauchi, Y. (1998). *Yami's origami: First steps to a thousand paper cranes*. Woodbridge, NJ: Woodbridge.

Yount, L. (1996). *Twentieth century women scientists.* New York: Facts on File.

Research References

Arhar, J. (1997). The effects of interdisciplinary teaming on students and teachers. In J. L. Irvin (Ed.), *What current research says to the middle level practitioner* (pp. 49–56). Columbus, OH: National Middle School Association.

Avery, C., & Avery, K. (2001). Kids teaching kids. *Journal of Adolescent & Adult Literacy, 44*(5), 434–435.

Bailey, L. (2000). Integrated curriculum: What parents tell us about their children's experience. *The Educational Forum, 64*(3), 236–242.

Basista, B., & Mathews, S. (2002). Integrated science & mathematics professional development programs. *School Science and Mathematics, 102*(7), 359–370.

Beane, J. (1993). *The middle school curriculum: From rhetoric to reality* (2nd ed.). Columbus, OH: National Middle School Association.

Beane, J. (1995). Curriculum integration and the disciplines of knowledge. *Phi Delta Kappan, 76*(8), 616–622.

Beane, J. (1997). *Curriculum integration: Designing the core of democratic education*. New York: Teachers College Press.

Bickley-Green, C. (1995). Math and art curriculum integration: A post-modern foundation. *Studies in Art Education, 37*(1), 6–18.

Bransford, J., Catterall, J., Deasy, R., Goren, P., Harman, A., Herbert, D., Levine, F., Seidel, S., & Sroufe, G. (2004). *The arts and education: New opportunities for research*. Washington, DC: Arts Education Partnership.

Bruner, J. (1977). *The process of education*. Cambridge, MA: Harvard University Press.

Burns, M. (1995). *Math and literature (K–3), vol. 1*. Sausalito, CA: Math Solutions.

Burns, M., & Sheffield, S. (2004). *Math and literature*. Sausalito, CA: Math Solutions.

Butzow, C., & Butzow, J. (2006). *The world of work through children's literature: An integrated approach*. Greenwood Village, CO: Teacher Ideas Press.

Capraro, R. M., & Capraro, M. M. (2006). Are you really going to read us a story? Learning geometry through children's mathematics literature. *Reading Psychology, 27*(1), 21–36.

Carr, K., Buchanan, D., Wentz, J., Weiss, M., & Brant, K. (2001). Not just for the primary grades: A bibliography of picture books for secondary content teachers. *Journal of Adolescent & Adult Literacy, 45*(2), 146–153.

Caskey, M. (2001). A lingering question for middle school: What is the fate of integrated curriculum? *Childhood Education 78*(2), 97–99.

Caskey, M. M., & Johnston, J. H. (1996). Hard work ahead: Authentic curriculum under construction. *Schools in the Middle, 6*(2), 11–18.

Cobb, P. (2000). The importance of a situated view of learning to the design of research and instruction. In J. Boaler (Ed.), *Multiple perspectives on mathematics teaching and learning* (pp. 45–82). Westport, CT: Greenwood.

Cornett, C. (2003). *Creating meaning through literature and the arts*. Upper Saddle River, NJ: Merrill/Prentice Hall.

Davies, M. A. (1992). Are interdisciplinary units worthwhile? Ask students. In J. Lounsbury (Ed.), *Connecting the curriculum through interdisciplinary instruction* (pp. 37–41). Columbus, OH: National Middle School Association.

Deasy, R. J. (Ed.). (2002). *Critical links: Learning in the arts and student academic and social development*. Washington, DC: Arts Education Partnership.

Dewey, J. (1924). *Democracy and education: An introduction to the philosophy of education*. New York: Macmillan.

Dewey, J. (1933). *How we think*. Chicago: Regnery.

Donoghue, M. (2001). *Using literature activities to teach content areas to emergent readers*. Needham Heights, MA: Allyn & Bacon.

Drake, S. (1998). *Creating integrated curriculum: Proven ways to increase student learning*. Thousand Oaks, CA: Corwin Press.

Drake, S., & Burns, R. (2004). *Meeting standards through integrated curriculum*. Alexandria, VA: Association for Supervision and Curriculum Development.

Draper, R. (2002). School mathematics reform, constructivism, and literacy: A case for literacy instruction in the reform-oriented math classroom. *Journal of Adolescent & Adult Literacy, 45*(6), 520–529.

Efland, A. (2002). *Art and cognition: Integrating the visual arts in the curriculum*. New York: Teachers College Press.

Eisner, E. (1998). Does experience in the arts boost academic achievement? *Arts Education, 51*(1), 5–15.

Eisner, E. (2004). *The arts and the creation of mind*. New Haven, CT: Yale University Press.

Fiske, E. B. (Ed.). (1999). *Champions of change: The impact of the arts on learning*. Washington, DC: Arts Education Partnership

Franke, M., & Kazemi, E. (2001). Teaching as learning within a community of practice: Characterizing generative growth. In T. Wood, B. Nelson, & J. Warfield (Eds.), *Beyond classical pedagogy in elementary mathematics: The nature of facilitative teaching* (pp. 47–74). Mahwah, NJ: Erlbaum.

Fredericks, A. (1991). *Social studies through children's literature: An integrated approach*. Greenwood Village, CO: Teacher Ideas Press.

Fredericks, A. (2000). *More social studies through children's literature: An integrated approach*. Greenwood Village, CO: Teacher Ideas Press.

Gallavan, N. (2001). Four lessons that integrate math and social studies. *Social Studies & the Young Learner, 13*(3), 25–28.

Gardner, H. (1997). A primer of multiple intelligences. *NEA Today, 15*(7), 17.

Gelineau, R. (2003). *Integrating arts across the elementary school curriculum*. Belmont, CA: Thomson Wadsworth

Grossman, P., Wineburg, S., & Woolworth, S. (2001). Toward a theory of teacher community. *Teachers College Record, 103*(6), 942–1012.

Hellwig, S., Monroe, E. E., & Jacobs, J. S. (2000). Making informed choices: Selecting children's trade books for mathematics instruction. *Teaching Children Mathematics, 7*, 138–143.

Hindin, A., Morocco, C., Mott, E., & Aguilar, C. (2007). More than just a group: Teacher collaboration and learning in the workplace. *Teachers and Teaching: Theory into Practice, 13(*4), 349–376.

Howey, K. (1996). Designing coherent and effective teacher education programs. In J. Sikula, T. J. Buttery, & E. Guyton (Eds.), *Handbook of research on teacher education* (2nd ed., pp. 143–170). New York: Simon & Schuster.

Hunsader, P. (2004). Mathematics trade books: Establishing their value and assessing their quality. *The Reading Teacher, 7*(57), 618–629.

International Reading Association (IRA). (2006). Excellent reading teachers: A position statement of the International Reading Association. In R. D. Robinson (Ed.), *Issues and innovations in literacy education* (pp. 19–24). Newark, DE: International Reading Association.

Jacobs, H. (1989). The interdisciplinary concept model: A step-by-step approach for developing integrated units of study. In H. H. Jacobs (Ed.), *Interdisciplinary curriculum: Design and implementation* (pp. 53–65). Alexandria, VA: Association for Supervision and Curriculum Development.

Jensen, E. (2001). *Arts with the brain in mind*. Alexandria, VA: Association for Supervision and Curriculum Development.

Johnson, N., & Giorgis, C. (2001). Interacting with the curriculum. *The Reading Teacher, 55*(2), 204–213.

Kaser, S. (2001). Searching the heavens with children's literature: A design for teaching science. *Language Arts, 78*(4), 348–356.

Kazemi, E., & Franke, M. (2004). Teacher learning in mathematics: Using student work to promote collective inquiry. *Journal of Mathematics Teacher Education, 7*, 203–235.

Kim, M., Andrews, R., & Carr, D. (2004). Traditional versus integrated preservice teacher education curriculum. *Journal of Teacher Education, 55*(4), 341–356

Kleiman, G. (1991). Mathematics across the curriculum. *Educational Leadership, 49*(2), 48–51.

Lave, J., & Wenger, E. (1991). *Situated learning: Legitimate peripheral participation*. New York: Cambridge University Press.

Leitze, A. R. (1997). Connecting process problem solving to children's literature. *Teaching Children Mathematics, 3*, 398–405.

Leu, D. J., Castek, J., Henry, L. A., Coiro, J., & McMullan, M. (2004). The lessons that children teach us: Integrating children's literature and the new literacies of the internet. *The Reading Teacher, 57*(5), 496–503.

Lieberman, A., & Grolnick, M. (1997). Networks, reform, and the professional development of teachers. In A. Hargreaves (Ed.), *Rethinking educational change with heart and mind* (pp. 192–215). Alexandria, VA: Association for Supervision of Curriculum and Development.

Louis, K., & Marks, H. (1998). Does professional community affect the classroom? Teachers' work and student experiences in restructuring schools. *American Journal of Education, 106*(4), 532–575.

MacGregor, M., & Price, E. (1999). An exploration of aspects of language proficiency and algebra learning. *Journal for Research in Mathematics Education, 30*, 449–467.

Martinez, M., & McGee, L. (2000). Children's literature and reading instruction: Past, present, and future. *Reading Research Quarterly, 35*(1), 54–169.

McCoy, M. (2003, Spring/Summer). Language, math, social studies, and . . . worms? *Integrating the Early Childhood Curriculum, 2*, 3–8.

McDonald, N., & Fisher, D. (2006). *Teaching literacy through the arts*. New York: Guilford.

Meinbach, A., Fredericks, A., & Rothlein, L. (2000). *The complete guide to thematic units: Creating the integrated curriculum*. Norwood, MA: Christopher Gordon.

Monroe, E., & Livingston, N. (2002). It figures: Language and mathematics add up through children's literature. *The Dragon Lode, 20*(2), 37–41.

Moss, B. (2003). *Exploring the literature of fact: Children's nonfiction trade books in the elementary classroom*. New York: Guilford.

Moyer, P. (2000). Communicating mathematically: Children's literature as a natural connection. *The Reading Teacher, 54*, 246–255.

Music Educators National Conference (MENC). (1994). *National standards for arts education*. Reston, VA: MENC.

National Association for Core Curriculum. (2000). *A bibliography of research on the effectiveness of block-time, core, and interdisciplinary team teaching programs*. Kent, OH: National Association for Core Curriculum.

National Council for the Social Studies (NCSS). (1994). *Curriculum standards for the social studies*. Silver Spring, MD: NCSS.

National Council of Teachers of English (NCTE) & International Reading Association (IRA). (1996). *Standards for the English language arts*. Urbana, IL: NCTE.

National Council of Teachers of Mathematics (NCTM). (1989). *Curriculum and evaluation standards for school mathematics*. Reston, VA: NCTM.

National Council of Teachers of Mathematics (NCTM). (1991). *Professional standards for teaching mathematics*. Reston, VA: NCTM.

National Council of Teachers of Mathematics (NCTM). (2000). *Principles and standards for school mathematics*. Reston, VA: NCTM.

National Middle School Association (NMSA). (1995). *This we believe: Developmentally responsive middle level schools*. Columbus, OH: NMSA.

National Research Council (NRC). (1996). *National science education standards*. Washington, DC: National Academy Press.

Palincsar, A., Magnusson, S., Marano, N., Ford, D., & Brown, N. (1998). Designing a community of practice: Principles and practices of the GIsML community. *Teaching & Teacher Education, 14*(1), 5–19.

Perkins, D. (1989). Selecting fertile themes for integrated learning. In H. H. Jacobs (Ed.), *Interdisciplinary curriculum: Design and implementation* (pp. 67–76). Alexandria, VA: Association for Supervision and Curriculum Development.

Phillips, P., & Bickley-Green, C. (1998). Integrating art and mathematics. *Principal, 77*(4), 46–49.

Putnam, R., & Borko, H. (2000). What do new views of knowledge and thinking have to say about research on teacher learning? *Educational Researcher, 29*(1), 4–15.

Rose, M. (2000). Lessons: Social studies/math millennium mastery. *Instructor, 109*(5), 14.

Roth, W., & McGinn, M. (1998). Knowing, researching, and reporting science education: Lessons from science and technology studies. *Journal of Research in Science Teaching, 35*(2), 213–235.

Schiro, M. (1997). *Integrating children's literature and mathematics in the classroom: Children as meaning makers, problem solvers, and literary critics*. New York: Teachers College Press.

Schwartz, S., & Pollishuke, M. (2005). Planning an integrated curriculum. In K. Revington (Ed.), *Creating the dynamic classroom: A handbook for teachers* (pp. 44–69). Toronto: Pearson Education Canada.

Scripp, L. (2002). An overview of research on music and learning. In R. J. Deasy (Ed.), *Critical links: Learning in the arts and student academic and social development*. Washington, DC: Arts Education Partnership.

Thompson, D., & Holyoke, K. (2000, Summer). Using children's literature to link mathematics and social studies: A multicultural exploration with bread. *Trends & Issues: The Publication of the Florida Council for the Social Studies, 12*, 22–24.

Vars, G. (1996). Effects of interdisciplinary curriculum and instruction. In P. S. Hlebowitsh & W. G. Wraga (Eds.), *Annual review of research for school leaders* (pp. 147–164). Reston, VA: National Association of Secondary School Principals and Scholastic Publishing.

Vars, G. (1997). Effects of integrative curriculum and instruction. In J. E. Irvin (Ed.), *What current research says to the middle level practitioner* (pp. 179–186). Columbus, OH: National Middle School Association.

Vars, G., & Beane, J. (2000). Integrative curriculum in a standards-based world. ERIC Digest. (ERIC Document Reproduction Service No. ED 441618) [On-line]. Available online at http://www.ed.gov/databases/ERIC_Digests/ ed441618.html

Walling, D. (2005). *Visual knowing: Connecting art and ideas across the curriculum*. Thousand Oaks, CA: Corwin Press.

Ward, R. (2003). How much is a billion? A lot more than you think! *Arizona Reading Journal, 30*(1), 27–29.

Ward, R. (2004a). Looking for math in all the right places. *The California Reader, 38*(2), 58–65.

Ward, R. (2004b). K–8 preservice teachers author a mathematical piece of children's literature. *The California Reader, 38*(1), 24–30.

Ward, R. (2004c). K–8 preservice teachers' journey into the global village: Exploring real-world data using children's literature and technology. *Arizona Reading Journal, 31*(1), 43–47.

Ward, R. (2005). Using children's literature to inspire K–8 preservice teachers' future mathematics pedagogy. *The Reading Teacher, 59*(2), 132–143.

Ward, R. (2006a, Spring). Modeling effective pedagogical strategies for teaching mathematics. *The Charter Schools Resource Journal*, 1–9.

Ward, R. (2006b, January). One if by land; *three* if by sea? *Mathematics Teaching*, 20–21.

Ward, R., & Muller, D. (2006, September). Algebra and art. *Mathematics Teaching, 198*, 22–26.

Watts, S. (2004, Summer). Arts-infused summer school. *New Horizons for Learning Online Journal, 10*(3), Retrieved March 12, 2007, from http://www.newhorizons.org/strategies/arts/watts2.htm

Whitin, D., & Whitin, P. (1996). Fostering metaphorical thinking through children's literature. In P. C. Elliott (Ed.), *Communication in mathematics K–12 and beyond, 1996 yearbook of the National Council of Teachers of Mathematics* (pp. 60–65). Reston, VA: National Council of Teachers of Mathematics.

Whitin, D., & Whitin, P. (2004). *New visions for linking literature and mathematics*. Urbana, IL: National Council of Teachers of English.

Whitin, D., & Wilde, S. (1992). *Read any good math lately?* Portsmouth, NH: Heinemann.

Whitin, D., & Wilde, S. (1995). *It's the story that counts*. Portsmouth, NH: Heinemann.

Wortham, S. (1996). Bringing it all together. In S. C. Wortham (Ed.), *The integrated classroom: The assessment-curriculum link in early childhood education* (pp. 326–346). Englewood Cliffs, NJ: Prentice Hall.

Young, J. (2001). Why are we reading a book during math time? How mathematics and literature relate. *The Dragon Lode, 19*(2), 13–18.

Appendix
Assessment Tools and Rubrics

As described in the introduction to this book, this appendix includes several assessment tools and rubrics (see below) that a teacher might employ as a means to better assess students as they engage in the literature-based activities. Other helpful assessment resources are included in the "Assessment Resources References" section.

Observation Log

Name: ____

Date	Activity	Observed Behavior

Comments:

Observation Log

Name: ______________________________

Activity: ______________________________

Date: ________________

Objectives or Goals	Observed Behavior	Comments

Observation Log

Name: ____

Date	Activity	Participation in Tasks 0: Little or none 1: Engaged 2: Fully engaged	Participation in Discussions 0: Little or none 1: Engaged 2: Fully engaged	Collaboration with Team 0: Little or none 1: Average 2: Above Average	Comments

Analytic Scoring Scale

Understanding the problem	0: Complete misunderstanding of the problem 1: Part of the problem misunderstood or misinterpreted 2: Complete understanding of the problem
Planning a solution	0: No attempt, or totally inappropriate plan 1: Partially correct plan based on part of the problem being interpreted correctly 2: Plan could have led to a correct solution if implemented properly
Getting an answer	0: No answer, or wrong answer based on an inappropriate plan 1: Copying error; computational error; partial answer for a problem with multiple answers 2: Correct answer and correct label for the answer

(Source: Charles, Lester, & O'Daffer, 1987)

Inventory of Student's Mathematical Disposition

Name: ______________________________

	Date	Comments
Confident in using mathematics		
Flexible in doing mathematics		
Perseveres at mathematical tasks		
Shows curiosity in doing mathematics		
Reflects on their own thinking		
Values applications of mathematics		
Appreciates role of mathematics		

(Derived from Stenmark, 1991, p. 34)

Mathematical Disposition Checklist

	Student 1	Student 2	Student 3	Student 4
Confidence • Initiates questions • Is sure answers will be found • Helps others with problems • Other:				
Flexibility • Solves problems in more than one way • Changes opinion when given a convincing argument • Other:				
Perseverance				
Curiosity				
Reflective				
Appreciation for mathematics				

(Derived from Stenmark, 1991, p. 34)

Observation Log

Group members: ___________________________________

Activity title: _____________________________________

Name	Assigned Task	Task Completed? (Y or N)	Comments

Group Assessment

Group members: ______________________________

Activity title: ______________________________

Did your group . . .	☺	😐	☹
Listen			
Talk about the task			
Cooperate			
Finish the task			

What went well? ______________________________

____

____

What would you do differently? ______________________________

____

____

(Derived from Stenmark, 1991, p. 34)

Collaborative Report

Student Name: ______________________________ Date: __________

Teacher Name: ______________________________ Date: __________

Scale: 2 – Mastered fully   1 – Mastered partially   0 – Did not master

Criterion or Task	Student Rating	Teacher Rating

Student's comments: ____

____

Teacher's comments: ____

____

Sample Writing Prompts

- In your own words, explain the meaning of . . .
- The most important thing I learned in math class today (or this week) is . . .
- The most important thing to understand about *polygons* is . . .
  (*Note:* Change *polygons* to the concept explored.)
- I discovered that . . .
- Explain your reasoning about . . .
- I know my solution is correct because . . .
- I feel confident about my solution because . . .
- I am still uncertain about . . .
- Describe any instances during which you became stuck and how you became unstuck while solving the problem.
- Describe a real-world experience/connection to the mathematical concept you learned about today.
- Write a letter to a classmate who did not attend class today so that he or she will understand what you learned about.
- Draw a picture or diagram showing how the concepts you learned about today are connected.

(Derived from Stenmark, 1991, p. 34)

Index